Today's Space Elevator

Space Elevator Matures into the Galactic Harbour

Peter Swan, Ph.D.
Michael Fitzgerald

Prepared for the
International Space Elevator Consortium
Chief Architect's Office

Sept 2019

Today's Space Elevator

Published by Lulu.com

info@isec.org

978-0-359-93496-6

Cover Illustrations:
Front – with permission of Galactic Harbour Association.
Back – with permission of Michael Fitzgerald.

Printed in the United States of America

Preface

The Space Elevator is a Catalyst for Change!

There was a moment in time that I realized the baton had changed hands - across three generations. I was talking within a small but enthusiastic group of attendees at the International Space Development Conference in June 2019. On that stage there was generation "co-inventor" Jerome Pearson, generation "advancing concept" Michael Fitzgerald and generation "excited students" James Torla and Souvik Mukherjee. The "moment" was more than an assembly of young and old. It was also a portrait of the stewards of the Space Elevator revolution -- from Inventor to Developer to Innovators. James was working a college research project on how to get to Mars in 77 days from the Apex Anchor and Souvik (16 years old) was representing his high school from India. The excitement and enjoyment of that moment soon faded; but, later it was evident that a baton had been passed. The iconic linkage of the generations that were on stage during the National Space Society event and the publishing of a study report by the International Academy of Astronautics announcing that we were definitely on the Road to the Space Elevator Era is remarkable. Unmistakably, the Space Elevator as the catalyst of change has already occurred. The future of the Space Mosaic supporting transportation, trade, exploration, enterprise, and research requires the Space Elevator as its enabler.

We believe that the Space Elevator program has developed its concept to such a degree that it is ready to initiate engineering testing! This document illustrates the status of the Space Elevator project as of the Fall of 2019. "Show-me" is the methodology of this report.

The vision of the International Space Elevator Consortium (ISEC) is to have a world with inexpensive, safe, routine, environmentally friendly and efficient access to space for the benefit of all. As its name suggests, one of the principle elements of the ISEC action plan is to promote the development, construction and operation of a Space Elevator infrastructure as a revolutionary and efficient way of getting into space. This special report is to emphasize the fact that we are ready to proceed to engineering testing. In addition, it is recognized that this is not a space project -- it is a transportation infrastructure. In fact, there is much to be said for the statement:

"The Space Elevator will be the transportation story of the 21st Century."

Peter Swan, Ph.D.
President of ISEC
September 2019

Acknowledgements

Thanks must go to the members of the International Space Elevator Consortium, and other Space Elevator enthusiasts who have been dreaming "big" for years now. The phenomenal work accomplished over the last ten years has allowed the Space Elevator body of knowledge to increase exponentially. Alison Berman described why "we do it" in "The Motivating Power of a Massive Transformative Purpose." Singularity Hub, Nov 8, 2016.

> "... mind-blowing breakthroughs don't just happen. They take teams of bright and dedicated people chipping away at the problem day and night. They take a huge amount of motivation, toil, and at least a few failures. To solve our biggest problems, we need people to undertake big tasks."

Each of us knows that when the Space Elevator is operational, movement off-planet proceeds robustly. We know that we will have an impact on the future and move humanity on a positive vector with hope. This Massive Transformative "Moonshot" is rewarding to work on and is being accomplished because we believe.

Thanks are also due to those who have contributed, especially Dr. Cathy Swan who read every word several times as an editor translating engineering into English. And thanks to Adrian Nixon, John Knapman, Jerry Eddy, and Dennis Wright for their review of the document. We would also like to thank Nixene Ltd. for its insight into single crystal graphene and its help at understanding its potential as the Space Elevator tether material.

Well done Space Elevator team!

Executive Summary

The Space Elevator has developed significantly over the last ten years; and, indeed 2019 was a "break-out" year. The leap from thinking about future concepts to the realization that an aggressive proposal of early developmental activities must be initiated during the fall of 2019, highlights a phase change in Space Elevators with several elements:

- from Space Elevator to Galactic Harbour
- from wishing for a material for the tether to having one successfully tested
- from an immature plan to a preliminary positive assessment of each technology within each system segment
- from silent discussions in small groups to advocacy across the world.

Much of this realization occurred during the 2019 National Space Society's International Space Development Conference (NSS ISDC) in Washington D.C. The ISEC, and other Space Elevator thinkers, spoke of and supported four main themes concerning the status of the Space Elevator.

- **Theme One**: Space Elevators are closer than you think!
- **Theme Two:** Galactic Harbour is a part of this global and interplanetary transportation infrastructure
- **Theme Three:** Space elevator development has gone beyond a preliminary technology readiness assessment and is ready to enter initial engineering validation testing -- leading to establishment of needed capabilities.
- **Theme Four:** The magnitude of the Space Elevator Architecture demands that it be understood and supported by many.

In addition, discussions with an active audience of space enthusiasts at the conference lead to remarkable conclusions.

1. *The Space Elevator is Environmentally Friendly*
2. *The Space Elevator is the economic engine of the environmentally green planet of the 21st Century, and*
3. *A strong statement of this fact must be one of the principle messages spread around the world.*

Part of the "show me" aspect of this ISEC Report about the Space Elevator program status is contained within the background information summarized in several large appendices. This will enable readers to have easy access to several Space Elevator facts and conclusions. This set of reference materials will enable developers to immediately understand the current status and find significant Space Elevator resources, such as a lexicon and a list of professional references.

Beyond the engineering refinements occurring in parallel across the industry, there is one major step that can be initiated in the near term to significantly move the Space Elevator forward. The establishment of a Space Elevator Institute would focus many of the issues for discussions, research, analysis and recommendations. This Institute would be chartered to look at the newly forming strategic mosaic of space. We are no longer just going to space, we will be conducting exploration, research, military operations, trade and commercial enterprises. Not only will the Space Elevator Institute focus upon the transportation infrastructure, but it would also identify and study the vast array of questions still to be answered. The placement of the Space Elevator inside the strategic mosaic of space will ensure that the exploitation of tremendous new access to space will leverage the lessons of history and enhance the safety of the enterprise. The strategic mosaic of space is taking form; it is composed of trade, enterprise, research, exploration, and military protection. The ability of the Space Elevator to be a logistics giant will ensure that this movement off-planet will result in an economic engine on (and near) Earth. The codification of the engineering transportation infrastructure will solidify the segment to segment relationships and support the satisfaction of system level requirements in preparation of design activities. A second thrust would focus on investigations into such areas as funding approaches, relationships with supporting governments around the world, discovering its rightful place in the interplanetary support activities, and determining best approaches to develop the Enterprise Infrastructure across the Galactic Harbours. The ability to assess current and near term activities with an historical view will make this new concept of a strategic space mosaic even more enticing to future generations.

This book is organized as a "show me" document to demonstrate 16 years of advancements beyond Dr. Edwards' modern Space Elevator concept. The initial chapter will illustrate the "why" of Space Elevators while showing some basic evaluations of where we are and where we came from. The second chapter will explain what the Space Elevator and Galactic Harbours are with a current (Fall 2019) baseline architecture. Chapter three will show the Architectural approach along with systems engineering status with a baseline schedule. The fourth chapter will delve into each segment of the system and show the conclusions of a major study completed by the International Academy of Astronautics [Swan, 2019] along with results from several ISEC year-long studies. Chapter five will jump into the four themes and evaluate the engineering claims with respect to program status. These four themes represent the Space Elevator and Galactic Harbour status for the fall of 2019. Chapter Six will lay the foundation for the future. Significant information will also be available within the Appendices:

Appendix A: Frequently Asked Questions
Appendix B: Space Elevator Terminology
Appendix C: Summary of ISEC Studies
Appendix D: Summary of IAA Studies
Appendix E: Summary of ISEC Architectural Notes
Appendix F: Space Elevator References
Appendix G: ISEC Description

Table of Contents

Chapter 1: Introduction

1.0 Space Elevator Program

The need for a Space Elevator program was recognized during the early part of the 21st century which lead to the formation of the International Space Elevator Consortium. This activity supported engineering studies and technical conferences discussing various aspects of a future Space Elevator and then a Galactic Harbour. Over the 11 years of ISEC's life, the Space Elevator program went from concept to engineering reality, although very preliminary and fragile. The essential elements motivating the volunteers within this organization revolve around the conviction that mankind must become a space-faring society and the realization of how much the initiation of a Space Elevator will further the well-being of the Earth and its inhabitants. This hope for the future was reflected inside its own vision and mission.

- Vision: A world with inexpensive, safe, routine, and efficient access to space for the benefit of all mankind.
- Mission: The International Space Elevator Consortium (ISEC) promotes the development, construction and operation of a Space Elevator (SE).
- Infrastructure as a revolutionary and efficient way to space for all humanity. ISEC is made up of individuals and organizations from all around the world who share a vision of mankind in space.

One of the persistent questions that is asked deals with the availability of tether material. This report begins with early demonstrated evidence that the material for Space Elevator tethers is real and has been produced in the laboratory. The chosen material is single crystal graphene with several universities producing half meter long by one tenth meter wide single crystals with tensile strengths that could support the Space Elevator. Across the history of the Space Elevator, the need for a material for a Space Elevator tether is obvious. As the recognition that single crystal graphene can finally be said to handle the stresses required and can be grown in lengths appropriate for Space Elevator tethers, the last technological victory seems to have happened. We have a material that should be ready for us when we need it (in the late 20's). [note: there are two additional materials which could work, currently backups for Space Elevator tethers: single crystal boron-nitride and advanced Carbon Nanotubes] It would seem that now that the material is identified, the program should naturally jump ahead. Indeed, that is where we are:

The Space Elevator and Galactic Harbour
Concepts are ready for Prime Time

1.1 Mission Strengths

Why Space Elevators? This key question must be answered each time ISEC produces a book or report as we must encourage, enthrall, challenge, explain, and provide hope for our global community of supporters, stakeholders and those we seek to influence. To anyone who periodically looks up from their chair and searches the heavens for the future of humanity, it is obvious that we are moving off-planet in a major fashion, and in the near future. In addition to regular American, Russian and European space activities, the Chinese have landed a rover on the Moon and are planning a space station. The Indians have orbited a spacecraft around Mars and the Japanese have a module attached to the International Space Station (ISS). The National Aeronautics and Space Administration (NASA) Jet Propulsion Lab (JPL) have identified over 1,300 near-Earth asteroids that are relatively easy rendezvous from Earth. There are three companies investing in mining resources on asteroids while there are multiple companies preparing to create small habitats on the Moon. In addition, there is a rocket company (SpaceX) that plans on building a colony of greater than 10,000 people on Mars within its CEO's lifetime.

To ensure that these dreams are encouraged and made successful, there must be a change in the approach to travel within our solar system. The cost to orbit must become a very small part of the overall investment and the arena must support infrastructures that can be used multiple times, not thrown away each time they are used. When one looks at the concept of Space Elevators, the answer is obvious. The future of humanity's travel within our solar system requires a Space Elevator infrastructure that provides access to space that have the following strengths[1]:

- Routine [daily] access to space
- Revolutionarily inexpensive [<$100 per kg] to GEO and beyond
- Commercial development similar to bridge building
- Financial Numbers that are infrastructure enabling
- Permanent infrastructure [24/7/365/50 years]
- Multiple paths when infrastructure matures
- Massively re-usable, no consumption of fuels
- Environmentally sound/sustainable - will make Earth "greener"
- Safe and reliable [no shake, rattle and roll of rocket liftoff]
- Low risk lifting
- Low probability of creating orbital debris
- Redundant paths as multiple sets of Space Elevators become operational
- Massive loads per day [starts at 14 metric tons cargo loads]
- Opens up tremendous design opportunities for users
- Optimized for geostationary orbit altitude and beyond
- Does not leave debris in LEO
- Co-orbits with GEO systems for easy integration

---

[1] Swan, P., Raitt, Swan, Penny, Knapman. International Academy of Astronautics Study Report, Space Elevators: An Assessment of the Technological Feasibility and the Way Forward, Virginia Edition Publishing Company, Science Deck (2013) ISBN-13: 978-2917761311

The bottom line for Space Elevators, and the solar system, is that they open up humanity's hopes and needs to expand beyond the limited resources and environment of our own planet. A Space Elevator is the enabling infrastructure ensuring humanity's growth within and beyond our solar system. There are two main reasons why the human race needs Space Elevator infrastructures:

- Chemical rockets cannot get us to and beyond Low Earth Orbit (LEO) economically.
- Space Elevators will enable programs to make the Earth "greener" with daily environmental liftoffs as well as enabling space based solar power and other environmentally enhancing activities.

1.2 Themes

At this time in Space Elevator transportation infrastructure development, there are four themes that need to be emphasized and understood. Each of these strengthens the position that Space Elevators are ready to begin development. This book's approach should help developers to understand what is next and how to accomplish it through the leveraging of our strengths. During the International Space Development Conference in June of 2019, these themes were presented and discussed at length. The themes are presented here:

- **Theme One**: Space Elevators are closer than you think!
- **Theme Two:** Galactic Harbour is a part of this global and interplanetary transportation infrastructure
- **Theme Three:** Space elevator development has gone beyond a preliminary technology readiness assessment and is ready to enter initial engineering validation testing -- leading to establishment of needed capabilities.
- **Theme Four:** The magnitude of the Space Elevator Architecture demands that it be understood and supported by many.

In Chapter Five, these four themes for the summer of 2019 are expanded upon. Significant verification for each positive statement comes from finished study reports from ISEC and the International Academy of Astronautics. In parallel with the study results shown from these reports, there will be other assets to pursue for information. The appendices contain many further discussions and facts, to include a quick summary of ISEC Architectural Notes. In addition, one of the Appendices will answer "Frequently Asked Questions." Each of these documents will help the reader understand where the Space Elevator is today and where it must go in the future.

1.3 Interplanetary Mission Support - Mission Expansion

The Space Elevator is the Galactic Harbour, and an essential part of the global and interplanetary transportation infrastructure. In the community of off-planet movement with NASA's newest move to put boots on the Moon by 2024, Space Elevators must be part of the discussions. The key here is that daily, routine, inexpensive, massive movement of payloads towards the Moon and Mars is a strength of the Apex Anchor with high velocity and daily launch windows. Recent research has demonstrated the strengths of Space Elevators with respect to future missions to the Moon, Mars, asteroids and beyond. This research has shown that releasing from an Apex Anchor enables rapid transit times to Mars as low as 77 days while enabling releases every day of the year. The research at Arizona State University has determined that release from the Apex Anchor will enable daily, massive, safe and inexpensive support to humanities expansion off-planet. Can you imagine "bus schedules" for logistics support to the Moon and Mars? This becomes the transportation story of the 21st Century.

1.4 Status of Space Elevator Progress Towards "Show me" (Fall of 2019)

The Fall of 2019 will soon be recognized as a "Sea State Change" in the development of Space Elevators. ISEC has shown that the concept of Space Elevators has moved beyond Preliminary Technological Assessment. This surfaced after:
- ISEC produced eight year-long studies with resulting reports (Appendix E).
- The International Academy of Astronautics produced two study reports supporting the concept (Appendix D).
- The Obayashi Corporation conduced an independent study that focused upon humans on the Space Elevator and massive movement of space based solar satellites to GEO (see titles in bibliography in Appendix F).
- Internal ISEC assessments were provided within a series of Chief Architect's Notes. (see www.isec.org).
- The agendas of major international space agencies are aligning to target human presence and/or settlements on the Moon and Mars.

As such, the Space Elevator team believes it has laid out a baseline of engineering details resulting in this "Sea State Change" along with a recognition of the remarkable potential for new and innovative missions for interplanetary support. This naturally leads to the need for others to recognize our approach and successes. All this analysis was based upon the statement:

Show me the Numbers!
Yes, we can show you the numbers!

This requirement to "show the numbers" was met over the last ten years with constant analyses and specific study topics focusing experts on topics of interest. The following few paragraphs show the gestation of this statement and the proof that we are definitely moving past the Preliminary Technological Assessment of Space Elevators towards development and operations.

2000 - 2003: Brad Edwards set the stage for a transition from a "thought experiment[2]" to a solid engineering design. He took the Space Elevator community from Jerome Pearson's (1975) and Yuri Artsutanov's (1960) estimates of what a Space Elevator could be using with deployed tethers. Brad rejected NASA's conference results (1998) which included cryogenic electromagnetic engines and an estimate of possibility within "300 years." He took us to an engineering solution that showed numbers and provided estimated answers. He was still basing it upon the discovery of carbon nanotubes (CNT) and their promise, which have not, as yet, developed in the tensile strength arena - YET. He set the stage for the modern day Space Elevator while showing the numbers.

2008 - 2019: The Academicians of the International Academy of Astronautics (IAA) conducted a four year study with 40 space experts and published in 2013[3]. They:

- Established engineering solutions and concepts
- Targeted a tether tensile strength of 50 GPa (gigapascals) instead of 150 GPa, assuming a density of 1.3 g/cc.
- Shifted to solar power vs. ground lasers [however, left option open as an alternative]
- Broadened the possible tether material to include CNT and/or Boron Nitride nanotubes.

The second study looked at the "Road to the Space Elevator Era." [Swan, 2019] In this four year study with 30 global space experts, the conclusions were[4]:

- The Earth Port, Headquarters & Operations Center, and Tether Climbers are all buildable with today's available technologies and engineering expertise.
- The GEO (geosynchronous equatorial orbit) Node, GEO Region, and Apex Anchor technologies are understandable and not an issue for development.
- The tether material is the pacing item for the development of Space Elevators. Currently, there are three viable materials that could grow into the needed strong-enough and long-enough material for a Space Elevator: carbon nanotubes, boron nitride nanotubes and single crystals, and continuous growth graphene. The community waits for those materials to mature to the level that can be used as a Space Elevator tether 100,000 km long and strong enough to support its own weight plus multiple tether climbers against the pull of gravity. Recent investigations explored the possibility for making single crystal graphene by a continuous process using liquid metal. It seems highly possible that continuous single crystal graphene will be manufactured in the coming years and this material should be

[2] Einstein's concept of Thought Experiment as an approach to thinking about complex topics. https://en.wikipedia.org/wiki/Einstein%27s_thought_experiments

[3] Swan, P., Raitt, Swan, Penny, Knapman. International Academy of Astronautics Study Report, Space Elevators: An Assessment of the Technological Feasibility and the Way Forward, Virginia Edition Publishing Company, Science Deck (2013) ISBN-13: 978-2917761311

[4] Swan, P., David Raitt, John Knapman, Akira Tsuchida, Michael Fitzgerald, Yoji Ishikawa, Road to the Space Elevator Era, Virginia Edition Publishing Company, Science Deck (2019) ISBN-19: 978-0-9913370-3-3

considered going forward for Space Elevator tethers.

2010 - 2016: In parallel, the Obayashi Corporation invested its own money into Space Elevator designs for Japanese needs (2015)[5]:
- Their driving functions were the transport of people to GEO and space-based solar power satellites for Japanese energy needs.
- As such, they went back to 150 GPA tethers for safety and the larger demand in loads.
- They wanted people on Space Elevators [requires safety double cable system and stronger cables].
- Their initial location is the far western Pacific vs. center of Pacific.

Both of these later concepts (IAA & Obayashi) were building upon the ideas of Brad Edwards as.
- they were based upon engineering numbers and calculations, and
- they are doable with a material at high tensile strengths

2008 - 2019: Since 2008, there have been multiple study groups assessing current and future Space Elevator concepts. The ISEC list of study reports on all segments of the system [as well an Architect's Notes] filled in some significant shortfalls that were in the Brad Edwards study. The appendix lists these studies and summarizes their results. In addition, the Architectural Notes by ISEC's Chief Architect relate directly to successes during this formative development phase. The Galactic Harbour concept evolved and was discussed at many venues in 2017. This concept ties the Space Elevator transportation system into a concept where commercial enterprises flourish and enhance the overall business case for Space Elevators.

Year	Study Title
2020	Interplanetary Mission Support (in development)
2019	ISEC Outreach Program (in draft)
2018	Design Considerations for Multi-Stage Space Elevator
2017	Design Considerations for Space Elevator Modeling and Simulation
2016	Design Considerations for GEO Node and Apex Anchor
2015	Design Considerations for Earth Port
2014	Space Elevator Architectures and Roadmaps
2013	Design Considerations for the Tether Climber
2012	Space Elevator Concept of Operations
2010	Space Elevator Survivability and Space Debris Mitigation

2018 - present: We believe we have a material that will approach the 150 GPa tether requirement. Although the IAA/ISEC approach needs much less than the old requirement, the new material [Single Crystal Graphene] could be manufactured in long lengths as "single crystals.[6]" (130 GPa, with a density of 2.2g/cc). The minimum requirement is 84 GPa at this density. This "announcement" of a suitable material can not be made quite yet — there are a few more hurdles to beat before it can be

---

[5] Ishikawa, Yoji, The Space Elevator Construction Concept, Obayashi Corporation, 2013, IAC-13-D4.3.6.

[6] Nixon, Adrian, Update on Graphene as a Tether Material. 2019 International Space Elevator Conference, Seattle, 16-18 Aug 2019.

claimed; however, experts in the know on this material are confident that it can be matured into a tether material for Space Elevators. The estimate for long tether material strong enough for Space Elevators is the late 2030's. . A recent letter from Nixene Ltd. stated:

> *"Joint planning between ISEC and Nixene Ltd anticipates the development testing and deployment of the Space Elevator tether within the next decade or two at a cost of $30bn." (see letter at end of chapter)*

2019 - present: Our community exists because Brad Edwards showed it could be done — with engineering estimates and real numbers. We believe this "show me" approach helps us remain credible. During the past 19 years, the emphasis has been on how to build. We now have a vision that surpasses Brad Edwards'. He was serving Earth. We are expanding beyond Earth. That is why ISEC insists on saying ... that the Galactic Harbour is the transportation story of this century. Our vision:
- It is a Galactic Harbour which connects the vertical with the horizontal. It is a ride to anywhere and everywhere.
- It co-locates interplanetary exploration with interplanetary trade and mission support. The Galactic Harbour will ensure the interplanetary paradigm will be robust and exciting.
- It feeds and powers much of Earth; while, it offers clean power
- Mission support from Galactic Harbours become the enabling factor in humanity's expansion off-planet.
- The Galactic Harbour is the future of Space Elevators

Our expansive vision of Brad Edwards' baseline isn't just bigger, it is more inclusive and more demanding. The new vision impacts are to be felt by everyone -- from now on. In addition, the Space Elevator team feels that the development of a Space Elevator program is on an aggressive path to success. Already, the baton is being passed from the co-inventor (Jerome Pearson) to the Innovative Thinkers (ISEC members + others like JSEA) and now towards the students.

1.5   Establishing a Space Elevator Institute

The establishment of a Space Elevator Institute is the logical next step as the community believes it is ready to step out briskly towards development. The key mission for the Institute would be to consolidate the current designs into a program and investigate the arena in which the Space Elevator and Galactic Harbour can provide support. This Institute will be chartered to look at the emerging strategic mosaic of space, and to study the remaining unknowns of developing the Space Elevator. The global movement off-planet and the phenomenal capabilities around the world in space are coming together during the 20's and early 30's to change not only space activities, but the human condition on our planet. We are no longer just going to space, we will be conducting exploration, research, military operations, trade and commercial enterprises. Low cost access to space will revolutionize the conduct of business beyond the atmosphere. Entrepreneurs will rapidly expand into the void of space developing businesses no one has thought of before. Countries will be leveraging their space strengths to expand the hopes and dreams of their people. Explorers will no longer be test pilots exclusively, but reflect the population of Earth.

Because of all these remarkable "movements," not only will the Space Elevator Institute focus on the transportation infrastructure, but would also focus on the vast array of questions still to be answered. The placement of the Space Elevator inside the strategic mosaic of space will ensure that the exploitation of the tremendous new access to space will leverage the lessons of history and enhance the safety of the enterprise. On the engineering and development side, the codification of the engineering transportation infrastructure will solidify the segment to segment relationships and support the satisfaction of system level requirements in preparation for design activities. The second main thrust of the Institute will be to focus on investigations into such areas as funding approaches, relationships with supporting governments around the world, discovering its rightful place within interplanetary support activities, and determining the best approaches to develop the Enterprise Infrastructure across Galactic Harbours. The ability to assess current and near term activities with an historical view will make this new concept of a strategic space mosaic even more enticing to future generations. The Institute will focus on:

- Leading diverse teams in the investigation of very different topics
- Formulating frameworks for new businesses to develop along the Space Elevator
- Establishing cooperative and collaborative study forums to address critical questions dealing with the Space Elevator's future
- Communicate the ideas, conclusions and concerns of the Institute, and
- Advocate Space Elevator solutions for moving off-planet and mission support of so many known and a vast number of unknown missions in the future.

The Space Elevator Institute will be instrumental in transitioning across the generations and will ensure the baton is passed to the young and innovative thinkers.

1.6   Overview of Chapters

This book is organized around the concept of "show me" and 16 years of advancements beyond the Dr. Edwards' modern Space Elevator. The initial chapter illustrated the "why" of Space Elevators while showing some basic realizations of where we are and where we came from. Chapter Two will explain what the Space Elevator and the Galactic Harbours are with a current (Fall 2019) baseline architecture. Chapter three will show the Architectural approach along with the systems engineering status with a baseline schedule. Chapter four will delve into each segment of the system and show the conclusions of a major study completed by the International Academy of Astronautics [Swan, 2019] along with results form several ISEC year-long studies. Chapter five will jump into the four themes and justify the engineering claims with respect to program status. These four themes represent the Space Elevator and Galactic Harbour status for the fall of 2019. Chapter six will lay the foundation for the future. Significant information will also be available within the Appendix List:

1.7   Letter from Nixene Ltd in support of Tether Material

One of the developments in the last year or so is the availability of a material called Single Crystal Graphene that appears to be suitable for Space Elevator tethers. There is tremendous potential for this material in several uses; however, Nixene Ltd is also working on a space elevator tether.

"Nixene Ltd is a low profile start-up graphene manufacturing company based in Manchester, UK. It has developed a concept to make single crystal graphene by a high-speed continuous manufacturing process. The founding purpose of this company is to make the material for the space elevator tether." (see letter following)

Nobel prize winning discovery unlocks the potential of the Space Elevator

The Space Elevator

Launch a satellite into orbit and lower a tether down to the earth surface. Then climb up the tether into space. That is the basic concept of the space elevator. This might sound like science fiction, but NASA and the International Academy for Astronautics (IAA) have funded feasibility studies to show this can be made a practical reality [1] [2]. Everything can be done with today's technology apart from one element – the tether requires a material strong and light enough to support its own weight and that of the payload. Until recently the candidate material was carbon nanotubes but manufacturing this material in the quantities needed has proved a challenge too far.

The Nobel Prize Discovery: Graphene

In 2010 two scientists at the University of Manchester, UK won the Nobel prize for discovering and isolating a new material called graphene [3]. Graphene is a new form of carbon 200 times stronger than steel yet flexible and transparent. It is the world's best conductor of heat and electricity. It has the highest melting point of any known material and is non-toxic [4]. This led to the discovery of a whole new class of materials called 2 dimensional (2D) materials. ISEC staff have shown that graphene is strong enough and light enough to make the space elevator tether [5]

Graphene is in the laboratory and is being developed rapidly by commercial industry with great potential (also for space elevator tethers)

Like any new technology, graphene has suffered from a certain amount of overhyping in recent years. Graphene is emerging from the hype and real-world applications are starting to emerge. To navigate the emerging technology, you need to understand there are two important forms of graphene,

      Graphene Powder
      Single Crystal Graphene

Single Crystal Graphene is the form that will ultimately manufacture the space elevator tether.

Current graphene manufacturing: 1 – Graphene Powder

Graphene is starting to revolutionise manufacturing. It is currently made in powder form and the Ford Motor company is already making quieter, stronger cars in the USA. For example, the 2019 Ford Mustang is now graphene enhanced [6]. Graphene powder has been added to asphalt roads in Italy where it doubles the service life of the highway [7]. However, graphene powder cannot be used to make the space elevator tether.

Current graphene manufacturing: 2 – Single Crystal Graphene

Single crystal graphene is the term used to describe a large-scale sheet of graphene with no defects. Making this material as a single molecule of carbon at the macro scale was thought impossible until the Chinese made this in the laboratory in July 2017 [8]. Since then other laboratories in China, South Korea and the USA have made single crystal graphene.

Manufacturing space elevator tether quality graphene – Nixene®

Nixene Ltd is a low profile start-up graphene manufacturing company based in Manchester, UK. It has developed a concept to make single crystal graphene by a high-speed continuous manufacturing process. The founding purpose of this company is to make the material for the space elevator tether. This company is based at the Graphene Engineering Innovation Centre (GEIC) at Manchester, UK and is part of the ecosystem of academics and industrial companies centred around the Nobel Prize winning scientist who isolated graphene. Nixene Ltd is currently seeking investment for the proof of concept process for manufacturing continuous single crystal graphene. The aim is to make a completely new material from multi layered graphene that is already being referred to as 'Nixene®'

On the way to make space elevator tether quality material Nixene® has many other uses. It can make ultra-lightweight armour, thermal shielding for hypersonic vehicles, enable ultra-tall kilometres high buildings and also offer the potential to create data cables with physical security that make them quantum computer proof.

The International Space Elevator Consortium (ISEC) is in close contact with Nixene Ltd and the CEO is now a board member of ISEC. Joint planning between ISEC and Nixene Ltd anticipates the development testing and deployment of the space elevator tether within the next decade or two at a cost of $30bn. More focussed resources would probably accelerate this programme. However; with the success of Nixene Ltd., the investment from the commercial arena is projected to be intense with potential uses described earlier and such massive markets as flexible-touch-screen devices, elevator tether for ultra-tall buildings (several km high) aircraft wings and safety devices of all types. Space Elevator tether development will leverage this massive commercial investment over the next several years.

Adrian Nixon

CEO, Nixene Ltd
12th September 2019

References:

1. Edwards, Bradley Carl. "The NIAC Space Elevator Program". (2003) NASA Institute for Advanced Concepts
    http://www.niac.usra.edu/studies/521Edwards.html
    [Accessed 8 September 2019]
2. Swan, Peter A.; Raitt, David I.; Swan, Cathy W.; Penny, Robert E.; Knapman, John M. (2013). Space Elevators: An Assessment of the Technological Feasibility and the Way Forward. Virginia, US: International Academy of Astronautics. pp. 10–11, 207–208. ISBN 9782917761311.
3. The Nobel Prize in Physics 2010. NobelPrize.org. Nobel Media AB (2019) [online] Available at:
    https://www.nobelprize.org/prizes/physics/2010/summary/
    [Accessed 8 September 2019]
4. The Royal Swedish Academy of Sciences. Scientific background on the Nobel Prize in Physics 2010. Available at: https://www.nobelprize.org/uploads/2018/06/advanced-physicsprize2010.pdf
    [Accessed 8 September 2019]
5. Graphene and Space Elevators: An interview with Adrian Nixon, (2018) National Graphene Association. https://www.nationalgrapheneassociation.com/news/graphene-and-space-elevators-interview-with-adrian-nixon/
    [Accessed 8 September 2019]
6. Ford Develops Car Parts Made Out of Graphene,(2019) Assembly Magazine.
    https://www.assemblymag.com/articles/94540-ford-develops-car-parts-made-out-of-graphene
    [Accessed 8 September 2019]
7. Successful first road trial results for Directa Plus, (2019) National Graphene Association.
    https://www.nationalgrapheneassociation.com/news/successful-first-road-surface-trial-results-for-directa-plus/
    [Accessed 8 September 2019]
8. Large single-crystal graphene is possible, (2017) Phys.org.
    https://phys.org/news/2017-07-large-single-crystal-graphene.html
    [Accessed 8 September 2019]
9. New record on the growth of graphene single crystals (2019) Phys.org.
    https://phys.org/news/2019-04-growth-graphene-crystals.html
    [Accessed 8 September 2019]

Chapter 2 Galactic Harbour and Space Elevator Baselines

This report has a basic set of assumptions that will establish system level baselines for discussions. These are shown in this chapter with descriptions of a modern day Galactic Harbour and a Space Elevator.

2.1 Preamble

The International Space Elevator Consortium (ISEC) has the mission to promote Space Elevator development. This requires that an understanding of where the concept comes from and how far along the project is. This leads to understanding the key historical lessons learned and recognition of the people who have contributed along the path of progress. This report lays out the need, and suggested approaches, for understanding the Galactic Harbour and Space Elevator baselines.

The Galactic Harbour and Space Elevator status must be established and strengthened. Identification of their strengths can be expressed in many ways; but, one of the quickest ways is to leverage past studies and Architectural Notes. The International Space Elevator Consortium has produced many programmatic and engineering studies over the last ten years that are directly applicable. The International Academy of Astronautics has produced two four-year study reports that emphasize that the "Space Elevator is Feasible" and that we are on the "Road to the Space Elevator Era. Each of these studies dealt with the future as the researchers saw it. Over the last five years, the Chief Architect for ISEC has been assessing various aspects of the development of Space Elevators and has recorded those findings in his Architectural Notes. The study reports and Notes are the basis for many of the conclusions and findings reflected within this book. The following few quick pages will help establish the basis for a Space Elevator and Galactic Harbour transportation infrastructure.

2.2 The Physical Soundness of the Space Elevator Concept

The foundation of any engineering project is a clear understanding of its physics. For the space elevator the two main questions to answer are:
* can it be built?
* will it stay up?

The answer to both is: Yes.

2.2.1 Physical feasibility
The feasibility of a tensile space elevator, given a strong enough tether material, has been known for decades[7][8][9][10]. The basic physics relies upon the balance of the Earthward

---

[7] Y. Artsutanov, "V Kosmos na Electrovoze," Komsomolskaya Pravda, 1960.

[8] J.D. Isaacs, A.C. Vine, H. Bradner and G.E. Bachus, "Satellite Elongation into a True Sky-Hook," Science 151, pp. 682-683 (1966).

gravitational force and the outward centrifugal force due to the Earth's rotation. How to build a structure that can support itself and climbers full of payloads using a minimum of material for a tether has a well-understood solution.

2.2.2 Stability

The space elevator is not a static structure. Its dynamic behavior while complex, is well-understood. An essential property of such a system is its stability. If the tether is disturbed in some way, will it return to its equilibrium state? A properly designed space elevator will do exactly that. This is due to gravity gradient stabilization[11] which is an effect of the inverse square law of gravity. If, for example, a space elevator is knocked one way or the other out of vertical, the resulting force on the tether is back towards the vertical. There will then be an overshoot of vertical, followed by a return again towards vertical. This sets up oscillations in the space elevator which can be dampened passively and/or actively. Understanding these oscillations is the main work of space elevator calculations and simulations. They arise from many forces such as; the motion of climbers on the tether, the non-sphericity of the Earth, the tidal effects of the Moon and Sun, the electromagnetic interaction of the tether with the magnetosphere, and the electromagnetic effects of solar storms. To date, many of these effects have been estimated; and none, so far, have posed a problem that cannot be solved or mitigated[12][13][14].

2.2.3 Conclusion

There are no essential physical uncertainties or obstacles to the construction of a space elevator. To be sure, unforeseen interactions of forces will arise as the project is studied in greater detail, but as for now, no insurmountable problems have arisen. The space elevator baseline presented in paragraph 2.5 is the result of a sound physical description of the forces and materials involved.

2.3   What is a Modern Day Galactic Harbour?

For the purpose of this book, a Space Elevator is a remarkable transportation infrastructure leveraging the rotation of the Earth to raise payloads from the Earth's surface into our solar system and beyond. It is indeed a part of a global transportation infrastructure. In a mature environment where Space Elevators are thriving in business and commerce, there would be several (probably up to ten) spread around the equator, each with a capability of lifting greater than 14 metric tons of payload per day, routinely and inexpensively. The Galactic Harbour will be the area encompassing the Earth Port [covering the ocean where incoming and

---

[9] J. Pearson, "The Orbital Tower: a spacecraft launcher using the Earth's rotational energy," Acta Astronautica 2, pp. 785-799

[10] B.C. Edwards and E.A. Westling, "The Space Elevator: A Revolutionary Earth-to-Space Transportation System," published by B.C. Edwards, Houston, Texas, 2003.

[11] V.V. Beletsky and E.M. Levin, "Dynamics of Space Tether Systems," p 21, published for the American Astronautical Society by Univelt, Inc., San Diego, California (1993).

[12] L. Perek, "Space Elevator: Stability," Acta Astronautica 62, pp. 514-520 (2008).

[13] S.S. Cohen and A.K. Misra, "The effect of climber transit on the space elevator dynamics," Acta Astronautica 64, pp. 538-553 (2009).

[14] A.M. Jorgensen and S.E. Patamia, "How Do Intense Magnetic Storms Affect a Space Elevator," 64th International Astronautical Congress, Beijing, paper # IAC-13-D4.3, 8X18785 (2013).

outgoing ships/helicopters and airplanes operate] and stretches, in a cylindrical shape, to include tethers and other aspects outwards towards its Apex Anchors.

In summary, customer product/payloads [satellites, people, resources, etc.] will enter the Galactic Harbour around the Earth Port and exit along the tether [to LEO (low earth orbit), GEO regions, Mars, Moon, asteroids, intergalactic, and towards the sun, dependent upon where it is released]. The "Galactic Harbour" is identified as the transportation "port" for the total transition from the ocean to release into space. The port will be three dimensional, not surface only. The concept is that payloads come into the Galactic Harbour. They are processed and released at some pier. The GEO Node is a good example of where a communications payload would be prepared for release, powered up, checked-out, and then released towards its assigned slot at GEO. The intra-transportation system is very similar to a train operation,

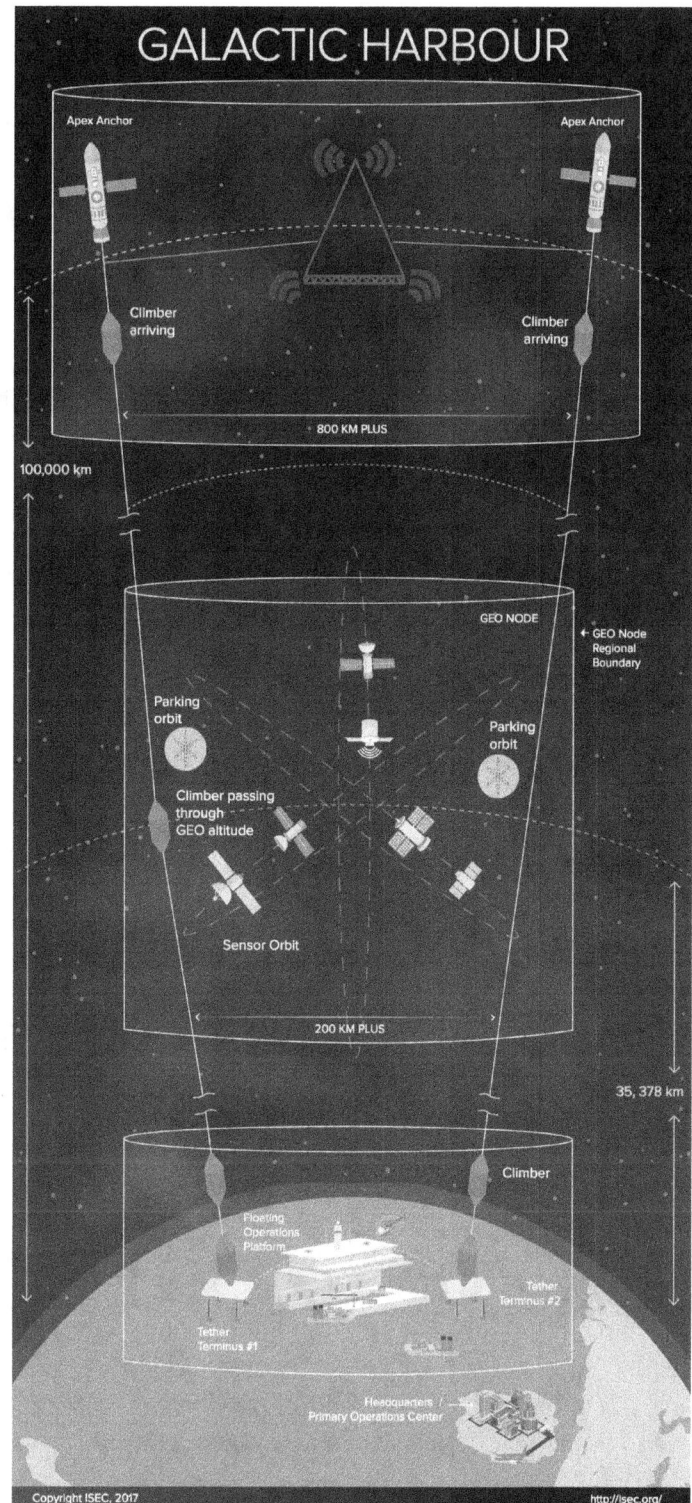

Figure 1 Galactic Harbour (2017)

movement on rails from one station (Port or Pier) to another. The difference is the Galactic Harbour will be up to 100,000 km high for payloads to be released at Apex Anchors.

15

The Galactic Harbour is the unification of Transportation and Enterprise[15]. As payloads start to move throughout Space Elevator systems, a core construction priority will drive businesses that will then lead to expansion beyond traditional functions. One projection is that the GEO Region will entice the construction of large enterprises to support non-traditional space businesses. A great representation of the advancements in the Galactic Harbour concept is shown in the following image of the Floating Operations Platform. This Galactic Harbour image shows the location for the transition from ocean going transportation infrastructure to vertical transportation infrastructure.

This copyrighted image of the Earth Port's Floating Operations Platform is based upon information presented in ISEC Position Paper #2015-1 entitled "Design Characteristics of a Space Elevator Earth Port" (ISBN 978-1-329-91060-7) and studies by the principals of Galactic Harbour Associates, Inc.

Figure 2, Floating Operations Platform
(with permission, Galactic Harbour Associates)

What one sees in the future are large commerce and industrial regions in space, supported by this new, revolutionary space access transportation system -- an elevator. A needed capability is the generation of power to be projected down to the surface of the Earth from geosynchronous altitude. Space based solar power systems will no longer be restricted by huge costs for access to orbit or restrained by rocket fairing sizes. Inexpensive delivery of payloads for construction purposes

---

[15] Fitzgerald, Michael, "Galactic Harbour, a Strategic Vision Emerges," Presentation at the National Space Society Conference, St. Louis, May 2017.

will lead to inexpensive power with almost zero carbon footprint on the surface of the Earth. Another mainline purpose will be to provide inexpensive access to all planets and moons in our solar system with routine release and capture enabled by the lack of a need for huge rockets and consumption of massive amounts of fuel. As the Space Elevator is built and deployed:

Galactic Harbours will unify transportation
and enterprise throughout the regions.

2.4 What is a Space Elevator

The elements of an historic Space Elevator system architecture (Figure 1) are considered to comprise of:

Earth Port: An ocean-going platform at the equator that supports movement of payloads to and from Space Elevator climbers.

Tether: A modern material that would extend from the surface of the ocean to an altitude of 100,000km. The material must be remarkably strong with a width and depth still to be determined (a width of about one meter with a depth of sub–micron are thought to be reasonable).

Tether Climber: The 'box' for the transportation of payloads. Current models suggest it will climb the tether using wheels with sufficient friction to move up/down as needed when supplied with power.

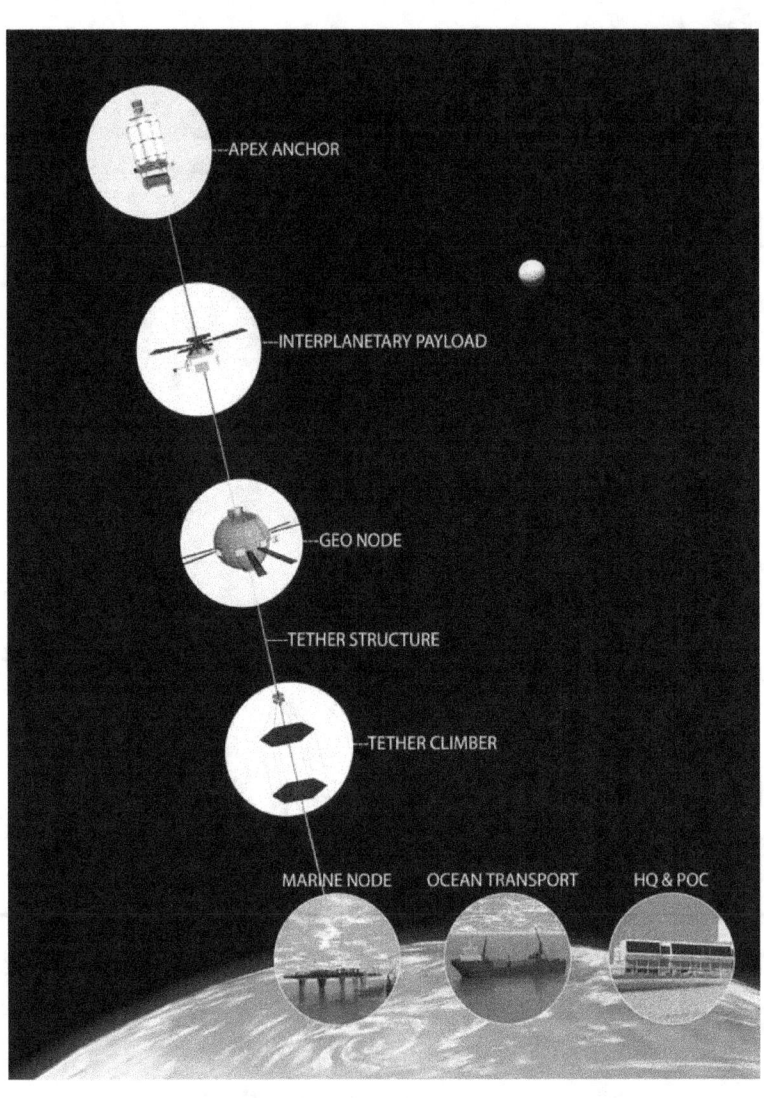

Figure 3: Space Elevator Architecture (2013)
(a Frank Chase image)

GEO Node: An altitude equivalent to modern day GEO satellites for off-loading payloads into this commercially significant orbit.

Apex Anchor Node: This would be the upper terminus at the high end (100,000 km altitude), and capable of on/off loading payloads. In addition, the Apex Anchor would be part of the system to control the dynamics of the ribbon. At this altitude,

18

the release velocity enables fast transit to the Moon, Mars and other solar system objects. Its value to interplanetary mission support will be revolutionary.

Headquarters/Primary Operations Center [HQ/POC]: This terrestrial location will be where the day-to-day operations occur for both Space Elevator transportation activities and business operations.

2.5   What is the Space Elevator Baseline?

When designing large complex space systems, there is a process which leads to engineering baselines that relate to customer needs. The Space Elevator Baseline will follow historic approaches within the space community.

> Systems can be defined as interoperating parts, pieces, components, subsystems, and/or segments with certain inputs, internal processes, and outputs intended to accomplish a given objective or set of objectives. To manage these independent entities as an operational system, it has become a common practice to identify requirements for each piece-part based on the operating concept of the system and its overarching architectural framework. As these requirements are established at the highest level of the system and allocated down to the appropriate segments, subsystems, or components, it is a good practice to establish them as a requirement or technical baseline.[16]

At this point in the development, the Transportation Baseline for the Galactic Harbour is:
- One Earth Port (a Floating Operations Platform and two Tether Termini).
- One GEO Region enabling multiple mission satellites to operate safely.
- One Apex Region with an Apex Anchor at the end of each tether.
- One Headquarters and Primary Operations Center (a major portion of which resides at the Earth Port FOP)
- Two tethers
- Operating Tether Climbers (estimates of seven per tether simultaneously), and
- Three Adjunct Elements to support the overall architecture
  - Space Debris Adjunct (mitigation and prevention)
  - Space and Surface Object Adjunct for situational awareness
  - Client Support and Management Adjunct

The Enterprise Baseline is not yet defined, but will be composed of many space based enterprises such as space based power, spacecraft assembly and testing, in-situ repair and refurbishment of space systems, as well a refueling on-orbit. These business activities are the primary initial customers and clients for the

---

[16] Willcox, TG, ESTABLISHING A PRODUCT BASELINE FOR GLOBAL POSITIONING SYSTEM SATELLITES THROUGH FUNCTIONAL AND PHYSICAL CONFIGURATION AUDITS, Masters Thesis, Naval Postgraduate School, 2011.

Transportation System; and as such, are the real source of transportation requirements. The Galactic Harbour is a Transportation project with multiple Space Elevators which will be supporting a growing set of customers and clients over time. (such as, GEO region enterprises, interplanetary enterprises, research activities, and more.) It is currently estimated that the GEO Region will become the busiest segment as it will host loading/unloading of climbers, providing power and communications to all enterprise activities, ensuring safe operations, and providing the monitoring and controlling for safety within the region. A representation of this activity at GEO is shown in Figure 4.

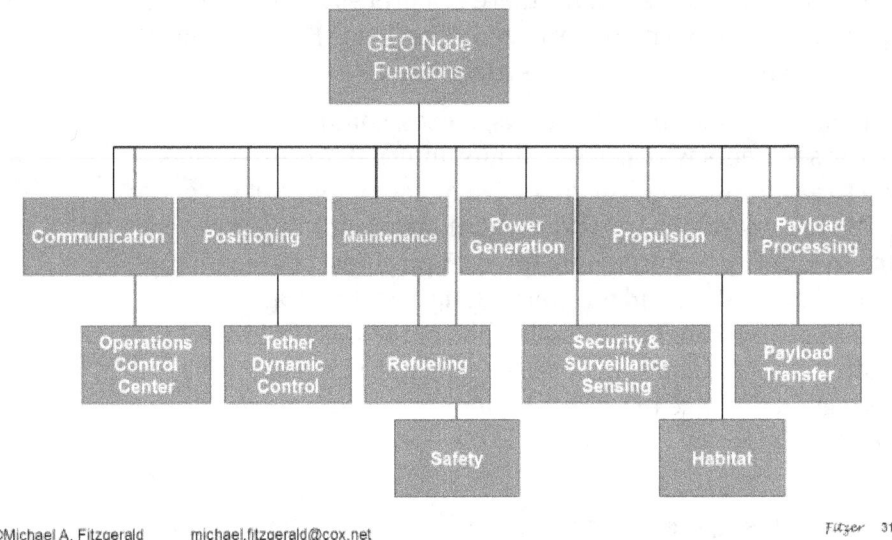

Figure 4, Geosynchronous Region Activities

Chapter 3: Architectural Approach

3.1 Introduction

Chapter Three shows the essence of the ISEC developmental program which stems from the belief that the Space Elevator community is moving from NOW to THEN and, of course, BEYOND. We envision moving from today – holding a vision of the Space Elevator; to initial operations – a marvelously engineered space transportation system; and to the full capability – to a gloriously robust enterprise within Galactic Harbours. The reader will first be shown the Strategic Approach as this sets the stage for any progress beyond today. The development of a plan of action can only be accomplished within a strategic approach with consensus among the significant developmental constituents. The next major topic for understanding the heart of our technology maturation and engineering validation processes is Sequences. This process is the source of our technical and intellectual fuel. Many of the early steps will be repeated until we "get it" and repeated until we "get it right" -- the essential definition of our perseverance. The next discussion will address the milestones to be successfully reached. It documents, in summary form, how we have entered the Space Elevator Era. The discussion explains how we moved from a roadmap study in 2014 to today when we are ready to declare "Tech Ready."

During the last 16 years, since Dr. Edwards' modern day design of a Space Elevator, there have been many approaches to proceed from concept to operations. The International Space Elevator Consortium has conducted many studies on the topic with a growing body of knowledge for this revolutionary transportation infrastructure. As a mega-project, the basic developmental steps will evolve; however, as it is a unique structure, there will be innovative concepts along the way. While there are many similarities with other mega projects, the differences are significant. This chapter will discuss the developmental process as it applies to the Space Elevator and lay out a proposed sequence of major milestones that are recognizable, measurable, and critical. A common understanding should flow through a consensus of ideas from the ISEC's body of knowledge leading to a recognition of similarities and differences with other mega projects.

3.2   Space Architecture Starts with a Strategic Approach

The reader knows that ISEC has a "Strategic Approach" for the development of Space Elevator. The community has spent much time discussing how to turn a long-term vision into a long-term "plan." The problem is that a plan usually implies either a specific schedule or a specific budget -- usually both. The Chief Architect has settled on the notion of "an approach;" disdaining the budget and schedule specifics for now. How much and when are exigencies -- the approach is immutable. Much of the following material was derived from the ISEC Architectural Note #9 as a portion of the architectural approach chosen.

3.2.1 Overall Strategy for International Space Elevator Consortium

**Our "strategy" is to link the Space Elevator Transportation
System to the Space Elevator Enterprise System
within a unifying vision: ... the Galactic Harbour.**

Why is there a need for an approach? For the most part, all of us agree that a Space Elevator will be a transforming transportation project of this century. With it, we can become a space faring people and support the planet with resources, energy, and so much more. In the International Academy of Astronautics Study #3.24 (jointly being authored by ISEC and the Japanese Space Elevator team), ten major categories of space endeavors are enumerated -- all enabled after Space Elevators start working.[17] In order to get Space Elevators working as a transportation system we need an approach.

How is an approach; especially a strategic one, formed? Forming an approach is not easy. The notion of "herding cats" immediately comes to mind. Herding researchers, scientists and professors is worse. However, during 2016's brainstorming session at our Seattle ISEC Conference, the attendees were convinced that we needed some order in the chaos. With some prodding, the team realized that the brainstorming participants saw the difference between the elevator and the business done near it and because of it. We had a first level of agreement and foresaw a "Space Elevator Transportation System".

After that conference, business and service functions would become part of a larger whole. The brainstormers were, in effect, cautioning that we needed to be careful that a Space Elevator Transportation System was not a "bridge to nowhere." The lesson of 'bridge to nowhere' is that, although a bridge is a separately engineered entity, it must be built to service the locale in which it is located. It must also help or enable improvement to that locale. In our minds, that means portraying our future transportation system and the enabled businesses within a "Unifying Vision." The vision is unifying because, though our first chore is to build the transportation system, it must be built to service the coming industries. Further, the transportation system must be built in a way such that it merges with the entrepreneurial activity. Hence the Space Elevator Transportation System merges with Space Elevator Enterprises. This latter point is critical. The manifestation of the enterprise is that it is an outgrowth of the transportation system. The transportation system must be able to grow and become part of a thriving enterprise. They are separate; but, they cannot be segregated from each other.

Breaking this into small steps requires some technical delineation. The team agreed that a Space Elevator will be an enabling force in this century. The ISEC team will begin a technical base lining activity which addresses how to get to Initial Operations Capability [IOC] for both baselines of the Space Elevator systems. This should enable:

---

[17] Swan, P., David Raitt, John Knapman, Akira Tsuchida, Michael Fitzgerald, Yoji Ishikawa, Road to the Space Elevator Era, **Virginia Edition Publishing Company**, Science Deck (2019) ISBN-19: 978-0-9913370-3-3

a. The assignment of building two technical baselines to a small, technical, system engineering working group.
b. Delineation of one of the two baselines:
   i. Space Elevator Transportation System and
   ii. Space Elevator Enterprise
c. The outlining of a the Space Elevator Transportation System baseline and citing IOC as the first destination of that baseline. This activity has the highest ISEC system engineering priority.
d. The outlining of a Space Elevator Enterprise baseline and explanation of its IOC relative to the Space Elevator Transportation system. This activity must have some system engineering priority.

The ISEC leadership team introduced the concept of the Space Elevator in the context of a Galactic Harbour at the International Space Development Conference in Saint Louis (6-9 June 2018). At that conference, they:
a. Stated that a Galactic Harbour is like most any other harbor -- a place of interacting transportation, major commerce and business activities.
b. Noted the parallel with classic harbors: Los Angeles / New York / Hong Kong / Singapore, with the Port of Los Angeles cited specifically.
c. Identified that a classic harbor is a meeting place of two forms of transportation: sea transportation and land transportation in our case.
d. Presented the example of a quick overview of The Port of Los Angeles vs. the Space Elevator: sea faring meets space-faring.

ISEC sees building a Space Elevator Transportation System as its first responsibility. The Strategic Approach enforces this and avoids building a galactic bridge to nowhere.

3.2.2 Sequence of Development - Natural Approach

This section shows the projected engineering maturation of technical capabilities for a Space Elevator as it goes from concept to an operational infrastructure. The development of these projected sequences was accomplished during a one-year study by the International Space Elevator Consortium on the topic "Design Characteristics of the Space Elevator GEO Node, Apex Anchor and Communications Architecture[18]." Many of the words and developments derive from a paper by Peter Swan at the International Astronautical Congress in 2017 in Guadalajara.[19]

> Sequences - The orderly steps [events] we must go through
> to reach Initial Operations Capability and go beyond[20]

The ability to actually develop a Space Elevator infrastructure requires many steps sequentially building from the research and development phase through the single

---

[18] Penny, Robert "Skip", Design Considerations for Geo Node, Apex Anchor and Communications Architecture [on-going 2016-2017]

[19] Swan, P., Fitzgerald, M. "Space Elevator Development Sequence," IAC-16, paper and presentation, International Astronautical Congress, Guadalajara, Sept 2016,. IAC-16-D4.3.8.

[20] Fitzgerald, Michael. "Space Elevator Initial Operations Capability," Paper and presentation at 2016 International Space Elevator Space Elevator Conference, Seattle, 2016.

string deployment, IOC, and finally to FOC. Part of our thinking is that the Space Elevator will grow by adding functions and services.

Each sequence phase has its own sub-sequence -- perhaps uniquely so. For example, the single string sequence might have sub-sequences of:
- Entrance criteria review,
- Simulation of the test event,
- Risk reduction and test data collection validation,
- Execution of the sequence phase itself,
- Assessment of the performance data collected, and Exit criteria review.

In the Projected Milestone Sequence as developed during the 2016 ISEC study, the proposed sequences are Space Elevator Developmental Phases:
1. Pathfinder
2. Seed Tether
3. Single String Testing
4. Operational Testing
5. Limited Operational Capability (LOC)
6. Initial Operational Capability (IOC)
7. Capability On Ramps leading to FOC
8. Full Operational Capability (FOC)

Each sequential phases[21] is explained here:

Pathfinder – The pathfinder initial step is designed as an in-orbit flight demonstration of all possible sub-systems and elements of a Space Elevator. The elements and sub-systems could include engineering models or simulations of the Apex Anchor, Tether, GEO Node, Mars Gate, Marine Node, and Headquarters & Primary Control Center with communications elements in place. It is essential to note that this early pathfinder in-orbit experiment can be achieved using near-term technologies – i.e. the tether material need not be a full-up Carbon Nano Tube or single crystal graphene ribbon. It could perhaps be composed of Kevlar or beta material of some type).

Seed Tether – This step in development will be the basis for a feasible first step in building a Space Elevator – deployment. The estimate of the technological readiness (in about 2031) will project for a much less capable ribbon being deployed and captured by a "start-up" Earth Port. The seed tether would probably be close to 100,000 km long with an end mass acting as a counterweight. Immediately, the buildup of the ribbon will be initiated with small climbers adding tether material in order to strengthen the total system.

Single String Testing – In many ways Single String Testing is a specific version of Operational Testing. Single string tests are conducted when a selected set of functions for the Space Elevator – or one of the Space Elevator's segments - are

---

[21] Fitzgerald, Michael. "Space Elevator Initial Operations Capability," Paper and presentation at 2016 International Space Elevator Space Elevator Conference, Seattle, 2016.

aligned and operating. In early forms, single string testing could "simply" be an end-to-end simulation of a segment or even the entire architecture. Single string testing is largely investigative -- aiding engineering progress and maturation. The Single String tests will never be construed to be an operational capability; but, it is clearly a necessary step.

Operational Testing – Operational testing is that set of test events intended to validate that a system or segment performs as designed in an operational context. Generally speaking, the tests envisioned here are defined based upon the extension of the development specifications, system engineering approach, overall test plan, and other such documents. Operational testing of the Tether Segment will require a ribbon deployed at full length.

Limited Operational Capability (LOC) – The idea of Limited Operational Capability (LOC) is similar to the baseball concept of spring training. All aspects of the Architecture are included when the hardware has been operationally deployed. This phase is good for assessing whether operational personnel are knowledgeable and trained, that payload customers are aware and understand how this Space Elevator works for them, and operational instruction documents (checklists) are finalized and vetted with "real" operations and operators.

Initial Operational Capability (IOC) – One key point about the IOC is that system engineering competency is part of what IOC is. These "engineering competencies" – validated by execution of the sequenced events – are the functional requirements of the Space Elevator at IOC. The Space Elevator will function as designed and tested, with safety, with certainty, be well observed, and in communications contact with HQ/POC. The ISEC also sees the Space Elevator as a valued part of the space business enterprise in the latter part of this century: a useful – indeed valued – partner with a wide set of business entities. That business value relationship is part of what the IOC is. IOC is also an enabling step to points beyond. The ISEC sees an Initial Operational Capability for the Space Elevator composed of three valuable entities: 1) mature system engineering competency,   2) solid business value to investors and customers, and finally, 3) a foundation for future growth.

Capability On-Ramps Leading to FOC – The need for Space Elevator capability growth after IOC is obvious; but to be clear, the Space Elevator post-IOC on ramp activity will be a formal process by which we add more of the IOC functionalities, improved versions of the IOC functionalities, and new Space Elevator functionalities. Many see the on ramp as "primarily" the way we bring new missions aboard. A prime example of that is the consideration of when a second tether becomes operational. The sequence approach simplifies that consideration. The second tether will immediately follow the first.

Full Operational Capability (FOC) – The visionary aspect of the Architecture includes tourism, interplanetary travel staging, hospitals, factories, power generation and a multitude of operational support services. The Full Operational Capability vision of a Space Elevator will expand with time and be achieved by expansion via the more, better, or new paradigm. The basis of each expansion will be the engineering maturation achieved by progressing through the sequenced steps cited in this paper.

Conclusions: The Space Elevator Body of Knowledge needs to increase in its understanding of the growth of Space Elevators from a concept to operations. This concept was developed to explain the projected sequence of events. The development of a Space Elevator will be complex and take place over many years, but there has to be an initial projection of what that would look like.

3.3 Engineering Stages - Where are We?

This discussion addresses a milestone for all members of ISEC. It documents, in summary form, how we have entered the Space Elevator Era. The discussion explains how we moved from a roadmap study in 2014 to today -- ready to declare "Tech Ready." All readers should at least examine the graphics and understand them. Industry must now get involved and ISEC should help them. The majority of this discussion has come from ISEC's Architectural Note #24.

In the last six years, ISEC's Technology Maturation approach has melded with a better definition of a Space Elevator engineering solution. The 2014 publication of ISEC's "Architecture and Roadmap" Report removed the shroud of mystery and myth from the Elevator's scope and complexity. The elevator was no longer a mystery. "Design Consideration" documents published between 2013 and 2017 delineated the engineering approach for the Tether Climber, Earth Port, GEO Region, and Apex Anchor. An Architecture simulation tool was selected. The last technology hurdle - strong material for the tether – was conquered. Based upon this technology maturity, and its engineering momentum, we expect that before the middle of this century a Space Elevator Transportation System will be built and operating. Further, the engineering substance of the Space Elevator has solidified and

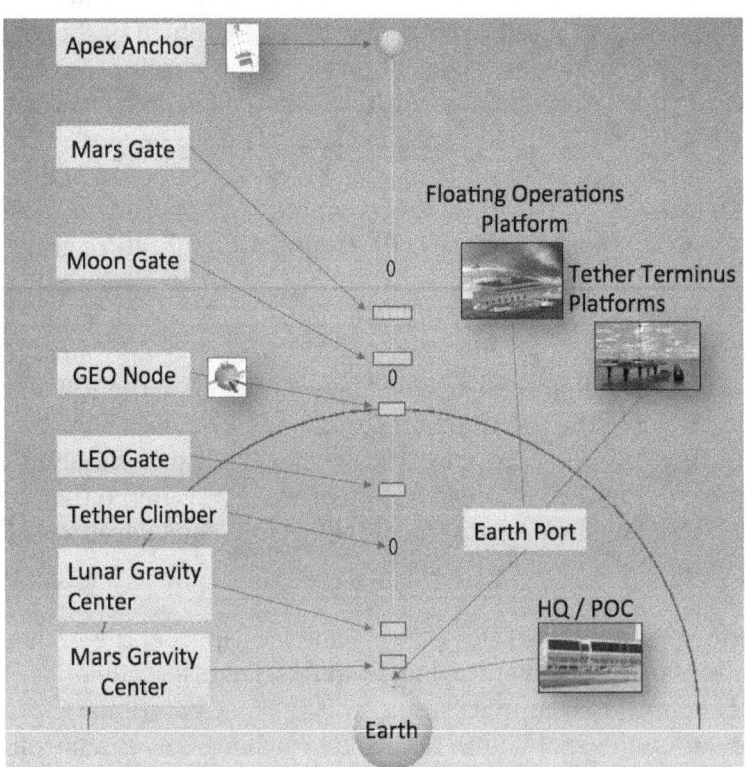

Figure 5, Space Elevator Description

become organized -- most notably as the Galactic Harbour. The Galactic Harbour will support enterprise activities along the GEO belt, factories and solar power generation near GEO, efficient interplanetary departures from the Apex and arrivals at GEO. Ultimately, products and materials will be delivered from space to the Earth Port. All this is closer than you think!

The technology momentum of the Galactic Harbour is real; and it underwrites the interplanetary vision of transportation, enterprise, and exploration. In the last year, the International Space Elevator Consortium asserted that the basic technologies needed are available; and, that each segment of the Space Elevator Transportation System is ready for engineering validation. The ISEC position has developed as follows:

1. The Galactic Harbour Earth Port is ready for an engineering validation program
2. An engineering validation program for the Space Elevator Headquarters / Primary Operations Center is ready to begin.
3. Tether Climbers development should start with engineering model assemblies, followed by an engineering validation program.
4. GEO Node engineering discussions and demonstrations can be accomplished with key members of industry and collaboration / outreach with certain government offices.
5. Apex Anchor is in the middle of engineering discussions and various simulations. Near term collaboration with engineering organizations and academia should be started.
6. Tether material is now "real." A prime material candidate has been identified. Production demonstrations are needed.
7. Collision avoidance needs have led to architecture engineering definitions which are being finalized. Candidate concepts have been identified. On orbit performance demonstrations are needed.

3.3.1 The Space Elevator is nearing the end of the Technology Development Phase

During the 2014 road mapping effort, it seemed evident that within the envisioned Space Elevator Architecture, new entities and technologies would be required, engineering approaches would need refinement, and new materials would need to be found. A technology development approach was built based upon a development approach of "Show Me." They were essentially a set of well-constructed demonstrations, inspections, tests, simulations, experiments and analyses - best conducted by industry (as industry will be building the elevator). ISEC also noted that each segment of the elevator has its own challenges and will likely need to resolve those challenges in segment unique manner. The technology and engineering issues facing something at the equator and in the middle of the Pacific Ocean are not directly relatable to something at the APEX Anchor at 100,000 kilometers above the equator in the middle of "Outer Space." As much as the issues are dissimilar; they are the same. They must be defined and their solutions found. After that it must be determined if something can be built from them. The work retains that theme. ISEC's technology development follows a tried and true sequence. The approach extends the thinking of industry / commercial Technology Plans.

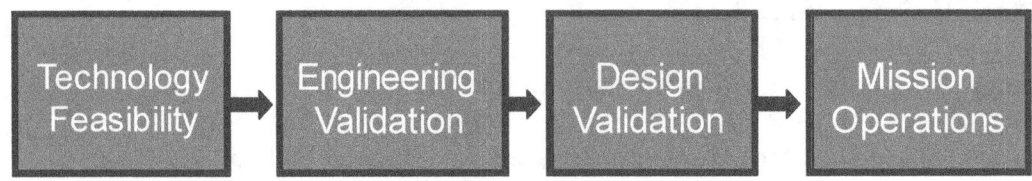

Figure 6, Engineering Developmental Phases

The progress within the plan continues to be based upon an iterative approach to risk mitigation. Recurring Technology Readiness Assessments culminating in operations demonstrations & prototypes such as success at Initial Operational Capability are required[22].

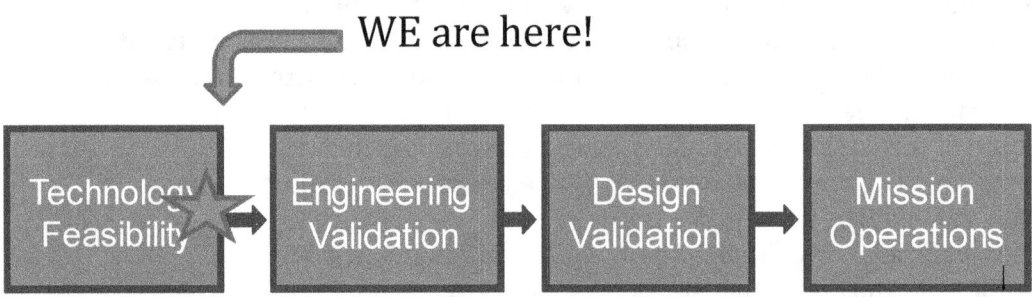

Figure 7, We are Here, Between Phases

The ISEC team has been assessing the technology feasibility situation since 2008. In recent times, the team has begun an open dialog with those members of industry, academia, and others; who could be the deliverers of ISEC solutions. Industry (especially) will show how the needed technologies are being matured and when they could be dependably available. These readiness assessments are the Phase One exit criteria:

- Document technology readiness state. Determine if the technologies are State of Art (SOA) or State of the Industry (SOI) or State of the Market (SOM)

- Establish readiness level rationale for all portions of the Program. Given that the technology availability has been demonstrated the level of readiness can be established for each program segment

- Set Success Criteria regarding Engineering Validation – the second phase. Prudent acquisition planning calls for an early design reviews. "Show me" means a lot at this point.

---

[22] Fitzgerald, Michael, Space Elevator Pathway to Technology Maturity ... and Beyond, From Fountains to Tech Ready. presented at 2019 International Space Elevator Conference, Seattle, 16-18 Aug 2019.

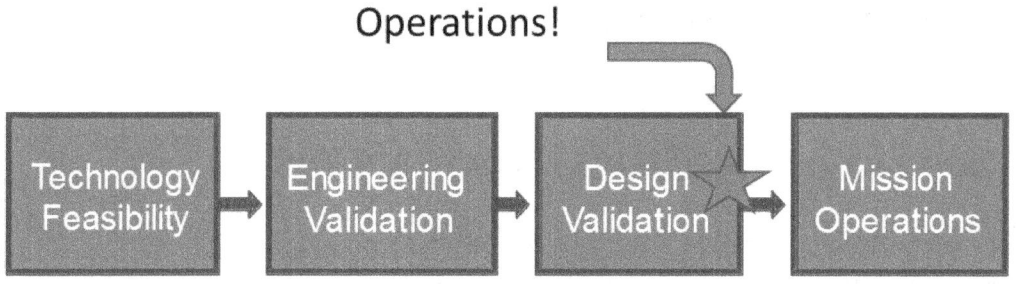

Figure 8, Operations is Downstream, the Fourth Phase

3.3.2 Phase two: Validate engineering approaches

Phase two will begin soon after phase one completion. Industry involvement is an imperative. Phase two activities are driven by six major activities:
- Examine Industry's production foundation
- Determine if the segments can be built
- Assess schedule & technical risk
- Delineate design criteria
- Set criteria and standards to enter the Design Validation Phase
- Baseline operations performance:

Chapter 4: Systems Engineering Status

4.1 Systems Engineering Status - Where are We?

The needs of each portion of a Space Elevator have been determined through a rigorous process that then lead to conclusions about each major segment of the architecture. The details were developed over the last few years by multiple teams around the world. The culmination of these "needs" and "conclusions" occurred during the writing of the report from a four year study[23] conducted by the International Academy of Astronautics. The members of the team were drawn from space experts who were either members of the IAA or were technical lead on the topic. Several universities, country-level administrations, multiple corporations, and two specific Space Elevator organizations (Japanese Space Elevator Association and International Space Elevator Consortium) contributed. This gathering of global experts across the technologies of a Space Elevator led to understanding of its overall needs. As these goals were approached systematically during the study, the results were presented in the form of needs and then followed by conclusions. When one looks at all the various technologies and where they are in the technology readiness level (TRL) evaluations common to NASA projects, the team results were stated within the study report for each major segment. These needs and conclusions are shown in this chapter as they are the key elements for the development of our conclusions that the Space Elevator is ready to proceed. The first example of these conclusions are shown for the overall Space Elevator Transportation Infrastructure.

- Overall Need #1 - Manage the immense Space Elevator domain. The Space Elevator covers a huge portion of "outer space." No other system or architecture has been this big. It will have dozens of moving parts and they will be moving in a variety of directions. The 'tyranny of distance" is going to take on new meaning.

- Overall Need #2 - Manage the operational interfaces between the Space Elevator and other space-based activities. The Space Elevator will not be alone.

- Overall Need #3 - Manage safety, surety, and recovery: The Space Elevator is something new. Managing its critical modes will get extensive attention during deployment and operational testing.

- Overall Need #4 - Develop and operate a simulation of the dynamic mass management model.

---

[23] Swan, P., David Raitt, John Knapman, Akira Tsuchida, Michael Fitzgerald, Yoji Ishikawa, Road to the Space Elevator Era, **Virginia Edition Publishing Company**, Science Deck (2019) ISBN-19: 978-0-9913370-3-3

- Overall Need #5 – Manage the human-rated functionalities of a Fully Operational Space Elevator.

Conclusions: The overall Space Elevator Transportation Infrastructure has technology momentum and should be taken beyond the Technology Readiness Phase. In the last year, the International Space Elevator Consortium asserted that the basic technology needs could be met, and each segment of the Space Elevator Transportation System was ready for engineering validation. Now each of the major segments of a Space Elevator will be discussed in a similar manner, beginning with the needs and followed by the conclusion.

4.1.1 Engineering Status of Earth Port

- Earth Port Need #1 – Development of a reel-in/reel-out system. The Earth Port Tether Terminus (Floating) Platforms will jointly and interactively operate the reel-in/reel-out system to maintain tether tension and stability. A tether dynamic simulation is an unquestioned need.

- Earth Port Need #2 – Development of climber to tether attachment. Once the dynamic forces acting at the Tether Termini, and the physical characteristics of the Earth Port end of the tethers are known, a working design of the climber(s) mechanism for attachment and detachment to the tether is needed.

- Earth Port Need #3 – Development of the climber power system to 40 km. The Earth Port will be responsible for providing power to the first segment of the climber's journey to the GEO Node and beyond. Numerous proposals for providing this power have been, and are still being, considered. In order to determine the technological requirements for the energy systems "aboard" the Earth Port's Floating Operations Platform and/or Tether Terminus Platforms, available solutions to this climber energy problem are needed.

- Earth Port Need #4 – Development of the Floating Operations and Tether Terminus Platform Structures. Today's technologies in Marine Architecture, Port Engineering and related fields can be employed to develop the structural and operational designs of the primary Earth Port facilities.

Conclusions: The Earth Port is buildable with today's available technologies and engineering expertise.

4.1.2 Engineering Status of HQ/POC

- HQ/POC Need #1 - The HQ/POC will coordinate all operations within the Space Elevator transportation infrastructure.

- HQ/POC Need #2 - Enable an embedded location monitor in all major devices inside the Galactic Harbour – probably GPS based -- transmitting to communications infrastructures at HQ/POC.

Conclusion: The Headquarters and Operations Centers are buildable today.

4.1.3 Engineering Status of Apex Anchor

The APEX Anchor represents the stabilizing element for the Space Elevator. When it gets to its station and settles into place, the Space Elevator will exist for the first time.

- Apex Anchor Need #1 - Development of the tether deployment satellite. The APEX Anchor will be born when the tether deployment satellite arrives. That satellite, especially the first one, will be a diverse, composite function satellite.
- Apex Anchor Need #2 - Development of the "reel-in/reel-out" system. The APEX Anchor and Earth Port will jointly and interactively operate the reel-in/reel-out system to maintain tether tension and stability
- Apex Anchor Need #3 – Develop the infrastructure allowing the orbital dynamics of parking service craft within the APEX Region.
- Apex Anchor Need #4 – Develop operational aspects and orbital dynamics of launches and landing from Interplanetary Gates.
- Apex Anchor Need #5 – Develop trust profiles and timing profiles to enable stabilization.
- Apex Anchor Need #6 – Develop a stabilization approach for initial tether deployment.

Conclusion: The Apex Anchor will be a challenge as its role is key to the building of the Space Elevator; but, it is not an engineering nor a technological issue.

4.1.4 Engineering Status of Tether Climber

- Tether Climber Need #1 - Interface with tether. The climber needs to grip the tether and enable climbing and descending along the full tether.

- Tether Climber Need #2 - Robotic space situational awareness.

Conclusion: The tether climber is so similar to a normal satellite design for today that there is no real technological or engineering challenge except for the interface with the tether itself. As there is a lack of information of the chosen material for the tether, some engineering must be resolved at a later time.

4.1.5 Engineering Status of Tether

- Tether Need #1 - Interface with tether climbers. The climber needs to be gripped to the tether and enable climbing and descending along the full tether.

- Tether Need #2 - Develop tether dynamic models for prediction of element location.

- Tether Need #3 - Develop methods to measure tensile strength and center of mass from initial deployment through Full Operational Capability.

- Tether Need #4 - Develop methods for perturbation control of the tether, including active damping from Earth Port, GEO Node, Apex Anchor and from moving tether climbers.

- Tether Need #5 - Develop approaches for tether repair.

- Tether Need #6 - Develop methods for bypass operations [one climber passing another], ie. spurs on a railroad track.

Conclusions: The tether material is the pacing item for the development of a Space Elevator. Currently, there are three materials that could grow into the needed strong-enough and long-enough material for a Space Elevator: carbon nanotubes, boron nitride nanotubes, and continuous growth graphene. The community waits for those materials to mature to the level that can be implemented into a Space Elevator tether 100,000km long and strong enough to support its own weight plus multiple tether climbers against the pull of gravity. At 100,000 kilometers long, a Space Elevator tether is a major engineering challenge. Recent investigations explored the possibility for making single crystal graphene by a continuous process using liquid metal. Making this a viable practical manufacturing process will be a significant effort over a period of years and probably many millions of dollars. However, such a process would create graphene products for many multi-billion dollar markets on the way to making the tether material. For this reason there is a credible return on investment case for manufacturing the material. This means it is highly possible that continuous single crystal graphene will be manufactured in the coming years and this material should be considered in any forward thinking about Space Elevator tethers.

Learning to extract best performance out of imperfect materials is a common engineering problem. Rare indeed is the design where all constraints and criteria are fully satisfied by a single solution. Two challenges that separate us from a current tether design become clear. The first, a challenge of assembly—how do we ensure uniform load distribution in our material, so that we can bring the nano-scale properties up to our macro-scale application. The second, a challenge of production—how to scale existing processes up to produce the volume needed. Neither of these challenges requires fundamentally new science or engineering. They require continued application of existing knowledge and skills. Based upon these conclusions, a number of recommendations can be made - the primary one is to encourage and support specific strength material development with the purpose of making them long enough and strong enough for Space Elevator tethers. [24]

4.1.6 Engineering Status of GEO Region

The GEO Node will be the transformational transportation system of the 21st

[24] Swan, P., David Raitt, John Knapman, Akira Tsuchida, Michael Fitzgerald, Yoji Ishikawa, Road to the Space Elevator Era, Virginia Edition Publishing Company, Science Deck (2019) ISBN-19: 978-0-9913370-3-3

century. It adds the third dimension to the world's logistical infrastructure. It will move objects, systems, material and (eventually) people from the Earth to Space. The Space Elevator will be vastly more efficient than today's launch systems.

- GEO Node Need #1 - Development of a tether deployment satellite. The APEX Anchor will be born when the tether deployment satellite arrives. That satellite, especially the first one, will be a diverse, composite function satellite.

- GEO Node Need #2 - Development of the reel-in/reel-out system. The APEX Anchor and the Earth Port will jointly and interactively operate the reel-in/reel-out system to maintain tether tension and stability

- GEO Node Need #3 - Robotic space situational awareness.

- GEO Node Need #4 - Develop the reel-in/reel-out system.

Conclusion: The GEO Node and GEO Region technologies are understandable and not an issue during development.

4.2   Preliminary Program Schedule

Over the last few years, including eight year long studies by ISEC and two four-year studies by the space experts at the International Academy of Astronautics, the consensus is that the Space Elevator is feasible and should be developed with a schedule similar to the one below.

Table 2, Space Elevator Proposed Schedule

Event Occurring between two dates	Early	Estimated
Material for Tether shows Characteristics	2019	2021
Material developed for Space Elevator Tether	2023	2029
Major Segments Validation Testing	2024	2030
Integrated Orbital Testing (Low Earth Orbit)	2031	2032
Launch of Deployment Satellite	2032	2034
Deployment of Space Elevator	2033	2036
Buildup of Space Elevator to Initial Operations Capability	2037	2040
Initial Operations	2037	2040
Galactic Harbour Operational	2037	2040
Second Galactic Harbour Operations	2039	2042
Full Operations Capability (with People)	2047	2057

Chapter 5: Space Elevator Program Status

5.1   Current Imperative!

Because of the availability of a new material as a potential solution for the Space Elevator tether material, the community strongly believes that a Space Elevator will be initiated in the near term. Indeed:

> *The Space Elevator and Galactic Harbour*
> *Concepts are ready for Prime Time*

5.2   Summer of 2019 Space Elevator Themes

The summer of 2019 was a turning point in the visibility of Space Elevator development and the future of movement off-Earth towards the Moon and the planets. As such, ISEC and other members of the Space Elevator community are active at the major conferences in Washington DC: National Space Society's International Space Development Conference (June 2019) and the International Astronautical Congress, sponsored by the International Astronautical Federation, International Academy of Astronautics, and International Institute of Space Law (Oct 2019).

This book is being developed to help in efforts to approach significant players in the space arena who are expected to be at the conferences and accessible to the ISEC team. The four themes to be presented are:

- **Theme One**:   Space Elevators are closer than you think!
- **Theme Two:**   Galactic Harbour is a part of this global and interplanetary transportation infrastructure
- **Theme Three:**   Space elevator development has gone beyond a preliminary technology readiness assessment and is ready to enter initial engineering validation testing -- leading to establishment of needed capabilities.
- **Theme Four:** The magnitude of the Space Elevator Architecture demands that it be understood and supported by many.

The following sections of this chapter illustrate each of the themes and provides background supporting information.

5.2.1 Theme One: Space Elevators are closer than you think

There are two major factors that have encouraged the Space Elevator community and have lad to this theme. The ISEC leadership believes that we will see a Space Elevator earlier than expected.

- Single Crystal Graphene is being developed and will be applicable for the Space Elevator tether. In the laboratory a 0.5x0.1 m sheet with 130 GPa tensile strength has been proven.
- The International Academy of Astronautics Study states:
  o The Earth Port, Headquarters & Operations Center, and Tether Climbers are all buildable with today's available technologies and engineering expertise
  o The GEO Node - GEO Region and Apex Anchor technologies are understandable and not an issue for development.

Recent investigations explored the possibility for making single crystal graphene by a continuous process using liquid metal. It seems highly possible that continuous single crystal graphene will be manufactured in the coming years and this material should be considered going forward for Space Elevator tethers.

5.2.2 Theme Two: Galactic Harbour is part of this global and interplanetary transportation infrastructure

With the latest revelations at the National Space Society's International Space Development Conference there are some remarkable aspects that are common across transportation infrastructures.
- The Space Elevator's Earth Port is the transportation nexus between Earth and the Solar System. Cargo and Payloads arriving by container destined for:
  – Geosynchronous enterprises
  – Interplanetary deliveries
- The Obayashi Corporation study (2015)25 designed a Space Elevator with:
  – People traveling to GEO, and
  – Space based solar power satellites for Japanese energy needs
- Release from the Apex Anchor enables interplanetary mission support in a robust manner. Recent studies at Arizona State University have shown that Apex Anchor releases could arrive at Mars in as little as 77 days with weekly "bus schedules" traveling in non-traditional Lambert method ellipses.

5.2.3 Theme Three: Space elevator development has gone beyond a preliminary technology readiness assessment and is ready to enter initial engineering validation testing -- leading to establishment of needed capabilities.

In the last six years, ISEC's technology maturation approach has melded with a better definition of Space Elevator engineering solutions. The 2014 publication of ISEC's "Architecture and Roadmap" report removed the shroud of mystery and myth from the elevator's scope and complexity. The Space Elevator was no longer a mystery. "Design Consideration" documents published between 2013 and 2017 delineated an engineering approach for Tether Climber, Earth Port, GEO Region, and Apex Anchor. An architectural simulation tool was selected. The last technology hurdle - strong material for the tether – will be overcome. Based upon this technological maturity, and its engineering momentum, we expect that before the

---

[25] Ishikawa, Yoji, The Space Elevator Construction Concept, Obayashi Corporation, 2013, IAC-13-D4.3.6.

middle of this century an operational Space Elevator Transportation System will be built and operating.

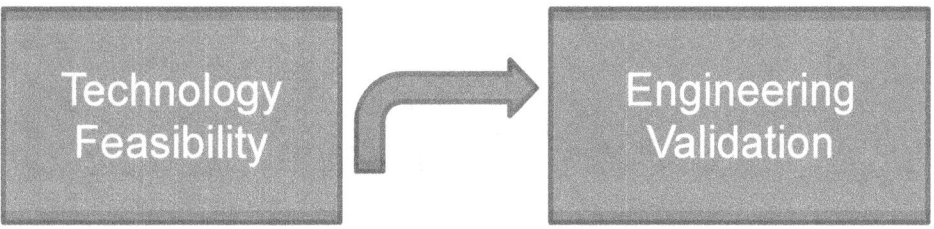

Figure 9, Technological Maturity as of Fall of 2019

Further, the engineering substance of a Space Elevator has solidified and become more organized -- most notably as the Galactic Harbour. The Galactic Harbour will support enterprise activities along the GEO belt, factories and solar power generation near GEO and efficient interplanetary departures from the Apex and arrivals at GEO.

The Technology Momentum of the Galactic Harbour is real; and, it underwrites the interplanetary vision of transportation, enterprise, and exploration

In the last year, the International Space Elevator Consortium assessed that basic technological needs are available, and each segment of the Space Elevator Transportation System is ready for engineering validation. The ISEC position:

1. The Galactic Harbour Earth Port ➔ ready for an engineering validation program
2. Space Elevator Headquarters / Primary Operations Center ➔ ready to start an engineering validation program
3. Tether Climber ➔ Engineering model assemblies needed -- then start an engineering validation program
4. GEO Node ➔ Engineering discussions and demonstrations with key members of industry are needed along with collaboration / outreach with certain government offices.
5. Apex Anchor ➔ Engineering discussions and various simulations are needed. Near term collaboration with engineering organizations and academia should begin follow-on outreach to key members of industry and government. Engineering validation follows.
6. Tether material ➔ Prime material candidate is identified; and, production demonstrations are needed.
7. Collision avoidance ➔ Architectural engineering definition is being finalized. Candidate concepts are identified. On orbit performance demonstrations are needed.

With all these thoughts, the preliminary technological readiness assessment is a process that the Space Elevator community, and especially ISEC, has embraced. As such the Space Elevator is ready to move into the validation testing Phase. The infrastructure is Ready to Proceed.

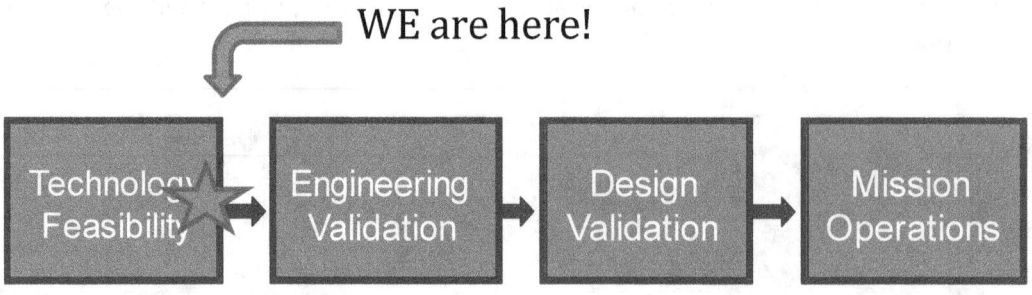

Figure 10, Space Elevator Level of Maturity

5.2.4 Theme Four: The magnitude of the Space Elevator Architecture demands that it be understood and supported by many.

There are several reasons why the Space Elevator Architecture needs to be included in broader discussions around the world because of the following two discussions:

Safe and reliable access to space is the foundation for humanity's travel within our solar system. The Space Elevator provides that access and enables:
- Routine [daily], Space Access
- Revolutionarily inexpensive [<$100 per kg] orbit transport
- Commercial development similar to bridge building [Public/Private]
- Financial Numbers that are infrastructure enabling
- Permanent infrastructure [24/7/365/50 years]
- Multiple paths when infrastructure matures
- Massively re-usable, no consumption of fuels
- Environmentally sound/sustainable - will make Earth "greener"
- Safe and reliable [no shake, rattle and roll of rocket liftoff]
- Low risk lifting
- Low probability of creating orbital debris
- Redundant paths as multiple sets of Space Elevators become operational
- Massive loads per day [starts at 14 metric tons cargo loads]
- Opens up tremendous design opportunities for users
- Optimized for geostationary orbit altitude and beyond
- Does not leave debris in LEO
- Co-orbits with GEO systems for easy integration

The Space Elevator is an invaluable addition enabling remarkable support of interplanetary missions because it not only supports Earth oriented satellites and missions, but it enables robust off-planet movement:
- Daily trips towards the Moon with roughly 14 hour transits
- Daily launches towards Mars with short transit times [as short as 77 days]
- High velocity releases from the Apex Anchor that can go to the outer planets with planetary gravity assists

5.3 Conclusion

The four "themes" chosen from the 2019 ISDC should be supported. Each of the four themes will have tremendous impact within the global transportation arena. The Space Elevator is ready for prime time. One constant realization is that ISEC needs to be invited, by space leaders, into discussions of significance. In addition, as discussed early in this report, the Space Elevator Institute should be created to provide more investigative power on issues of importance. The engineering refinements and the tie to business enterprises must be understood and executed.

Chapter 6: The Future of Galactic Harbours

6.1   Introduction

During the summer of 2019, the Space Elevator team recognized that it was a year for breakouts.   By recognizing that the Space Elevator should be invited into the global discussions on space and movement off-planet, remarkable collaborations will occur.   The Space Elevator must be included inside the newly developing strategic mosaic of space. This will ensure that the exploitation of the tremendous new arena of space will leverage the lessons of history and enhance the safety of the enterprise. The timing of Space Elevator development is tied closely with the amazing movement of the human race off-planet. Inexpensive and routine access to space will enable the people of Earth to have hope for an exciting future. Seeing humans residing on the Moon and Mars will bring home the value of Earth and all it means. The Space Elevator will enable this movement.

6.2   First Big Step

The realization that the Space Elevator is no longer a concept but an achievable mega-project will energize many people to support and implement its development. To help this happen, the creation of an institution would seem natural. The creation and support of a Space Elevator Institute will be the first major step in the refinement of the idea and the first move towards a developmental program. The establishment of the Institute will insure that major questions are investigated and discussions initiated to address complex relationships between new projects and government oversight. This newly developing mosaic of space will have complexity no one has thought of before. The Space Elevator Institute will address these issues early and bring together a consensus of thought to ensure the program can move forward. The two basic thrusts for the Institute would be:

Investigate the Transportation Baseline: This would include the codification of engineering concepts so that complexities can be reduced. The trades between various engineering options will be addressed and recommendations presented to ensure consistency within the program. Several of these include:
* Engineering concerns in the atmosphere
* Choice of material and tether approach
* Design of initial deployment satellite
* Develop an engineering simulation capability to represent Space Elevators

Investigate Engineering and Research Topics identified by the Space Elevator leadership: This activity would be "study" oriented where the complexity of development would be studied and analyzed with recommended paths as the product. There are many areas outside of strict engineering that must be assessed and recommended paths surface. Several of these include:
* Location of Earth Port

- Geosynchronous Orbit designation
- Space Debris communication and coordination
- GEO Region monitoring and coordination
- Quantify Interplanetary Mission Support needs
- Develop mission based simulations to represent Daily Operations

Many basic research topics have already been identified as essential to the development of certain segments of the Space Elevator system. These include:
- the mechanical, thermal and electromagnetic properties of the bulk tether material
- the effect of magnetospheric fields and solar radiation on tether motion and tether climbers
- the characterization of possible perturbations of tether motion and estimation of their effect relative to stable tether oscillations
- alternative types of tether-gripping mechanisms such as linear motors
- alternative types radiation protection such as active shielding.

The study of these topics and others will likely constitute the mandate for a research division within the Institute.

6.3   Space Elevator Institute Charter

The charter will be developed as the concept matures and the details surface. However, the mission is pretty straightforward. The Space Elevator Institute should have a mission such as:

Mission: Leverage the understanding of Space Elevators and Galactic Harbours to combine space and transportation futures into a common thread within the mosaic of space.

6.4   The Future

The beauty of this future for space transportation revolutions is that it is fast approaching. The Space Elevator community is rapidly arranging to meet it with enthusiasm and knowledge. Movement off-planet will demand low cost access to space. The Space Elevator will provide that with daily, routine, massive, and environmentally friendly infrastructure that will resemble "train operations." This infrastructure for transportation will lead to:

The Space Elevator will be the Transportation Story of the 21st Century

Appendices

Space Elevator (Frank Chase Image 2013)

Appendix A: Frequently Asked Questions

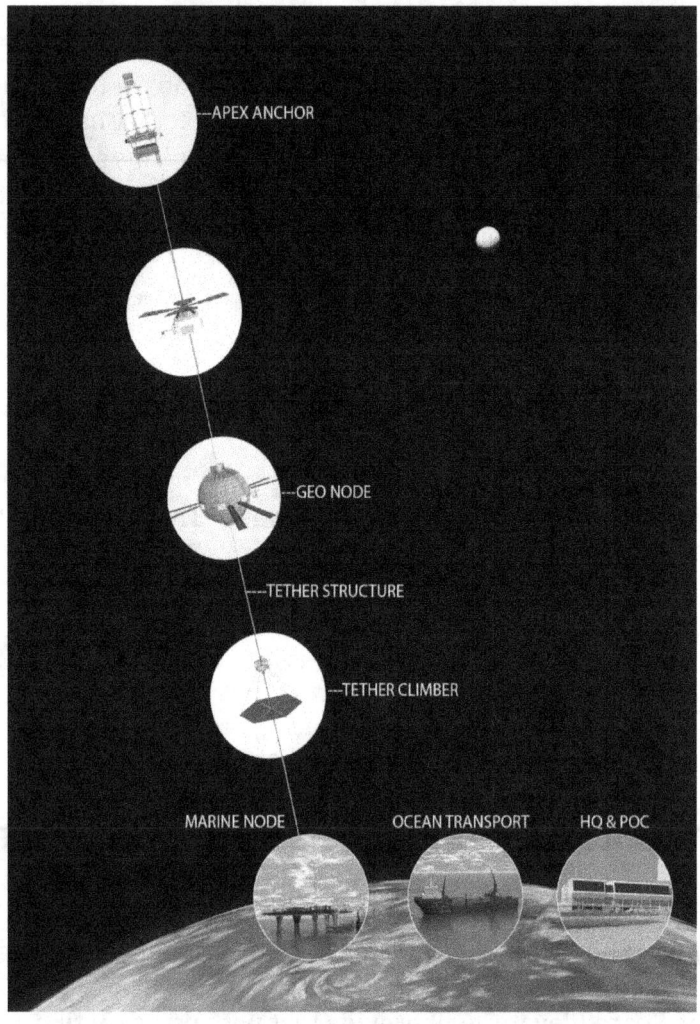

The following questions are answered below:

- *What is a Space Elevator?*
- *What is a Galactic Harbour?*
- *Why Space Elevators and Galactic Harbours?*
- *How will the Space Elevator Work?*
- *How will the Space Elevator Tether Stay Vertical?*
- *Who Invented the Space Elevator?*
- *How Strong does the Tether have to be?*
- *How will the Space Debris Environment be Handled?*

What is a Space Elevator? A Space Elevator (SE) can be thought of as a vertical railroad into space. A cable (Tether) stretches from the ground to an Apex Anchor (counterweight) 100,000 km up/out in space. Elevator cars (Climbers), powered by electricity travel up and down the Tether and carry cargo and eventually humans to and from space.   The Space Elevator is the most promising Space Transportation system on the drawing boards today, combining scalability, low cost, qualify of ride, and safety to deliver truly commercial-grade space access – practically comparable to a train ride to space.

Rocket-based space launch systems are inherently limited by the physics of rocket propulsion. More than 90% of the rocket's weight is propellant, and the rest is split between the weight of the fuel tank and the payload. It is very difficult (if not impossible) to make such a vehicle safe or low cost. A target cost of $1,000 US per kg is proving to be impossible to reach. In comparison, airliners charge us about $1 per pound, and train transportation is in cents per pound.

The Space Elevator is based on a thin vertical tether stretched from the ground to a mass far out in space, and electric vehicles (climbers) that drive up and down the tether. The rotation of the Earth keeps the tether taut and capable of supporting the climbers. The climbers travel at speeds comparable to a fast train, and carry no fuel on board – they are powered by a combination of sunlight and laser light projected from the ground. While the trip to space takes several days, climbers are launched once per day. The first "baseline" design will use 20 ton climbers, but by making the tether thicker (which can be done using the Space Elevator itself) we can grow the Space Elevator to lift 100, or even 1,000 tons at a time. In addition to launching payloads into orbit, the Space Elevator can also use its rotational motion to inject them into planetary transfer orbits – thus able to launch payloads to Mars, for example, once per day. Imagine the kind of infrastructure we can set up there, waiting for the first settlers to arrive… Looking back from the year 2100, the construction of the Space Elevator will be considered to mark the true beginning of the Space Age, much like the advent of the airplane or steamboat heralded the true commercial use of the air and sea.

What is a Galactic Harbour? For the purpose of this book, a space elevator is a tremendous transportation infrastructure leveraging the rotation of the Earth to raise payloads from the Earth's surface towards space and our solar system. It is indeed a part of the global transportation infrastructure. In a mature environment where space elevators are thriving in business and commerce, there would be several (probably up to ten) spread around the equator, each with a capability of lifting off greater than 20 metric tons of payload per day, routinely and inexpensively. The Galactic Harbour will be the area encompassing the Earth Port [covering the ocean where incoming and outgoing ships/helicopters and airplanes operate] and stretches up in a cylindrical shape to include tethers and other aspects outwards towards Apex Anchors. In summary, customer product/payloads [satellites, people, resources, etc.] will enter the Galactic Harbour around the Earth Port and exit someplace up the tether [to LEO, GEO regions, Mars, Moon, asteroids, intergalactic, and towards the sun, dependent upon where it is released]. The "Galactic Harbour" is identified to be the transportation "port" for the total transition from the ocean to release in space. The port would be three dimensional, not surface only. The concept is the payload comes into the Galactic Harbour. It is then processed and released at some pier. The GEO Node is a good example of where a communications payload would be prepared for release, powered up, checked-out, and then released to float towards its assigned slot at GEO. The intra-transportation is very similar to a train operation, movement on rails from one station (Port Pier) to another. The difference is the Galactic Harbour will be up to 100,000 km high for payloads to be released at Apex Anchors.

Figure 2 Galactic Harbour

The Galactic Harbour is the unification of Transportation and Enterprise. As payloads start to move throughout the space elevators, a core construction priority will drive businesses that will then lead to expansion beyond traditional functions. One projection is that the GEO Region will entice the construction of large enterprises to support non-traditional space businesses. What one sees now are a magnificent, large commerce and industrial regions in space, supported by this new, revolutionary space access transportation system; an elevator. A needed capability is the generation of power to be projected down to the surface of the Earth from GEO. This Space Based Solar Power will no longer be restricted by huge costs for access to the orbit. Inexpensive delivery of payloads to GEO for construction purposes will lead to inexpensive power with almost zero carbon footprint on the surface of the Earth. Another mainline purpose will be to provide an inexpensive access to all planets in our solar system (as well as our own Moon) with routine release and capture enabled by the lack of a need for huge rockets and consumption of massive amounts of fuel. As the space elevator is built and deployed, the:

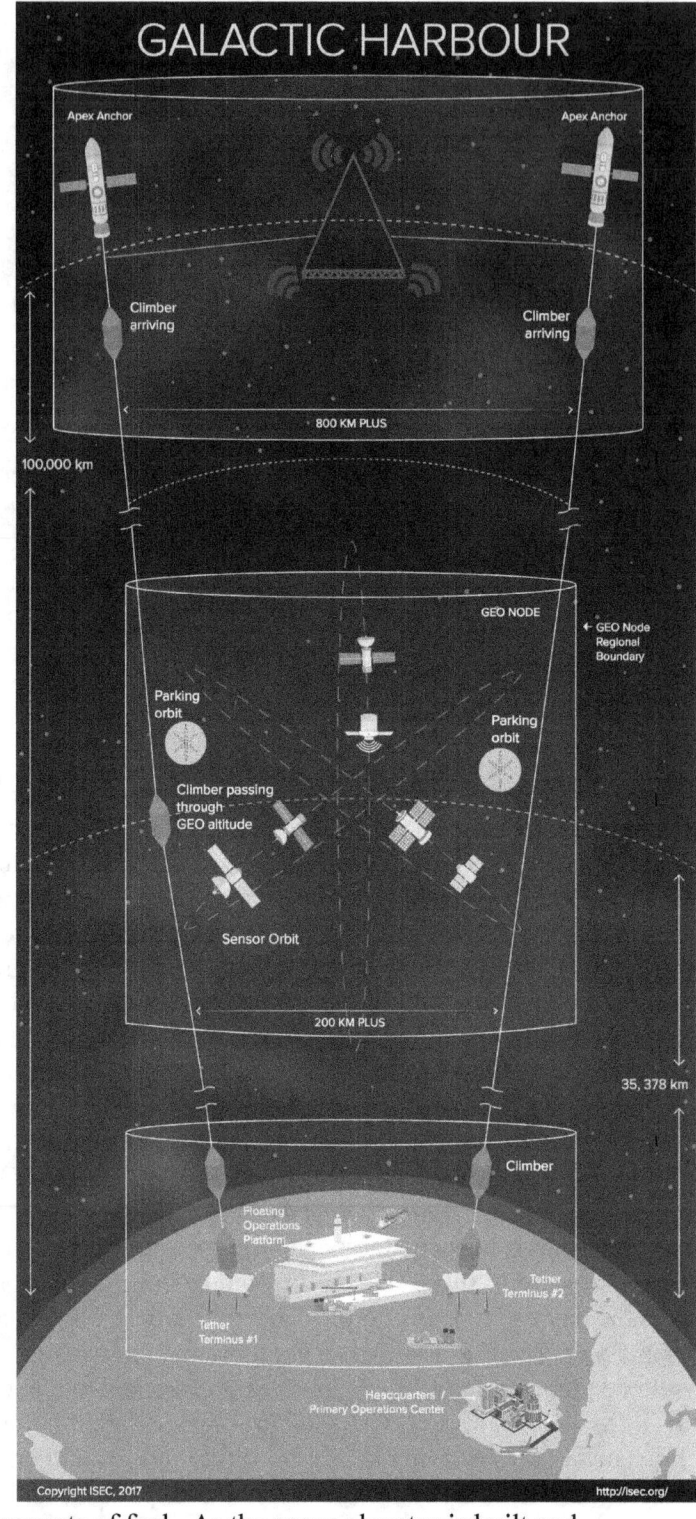

Galactic Harbours will Unify Transportation and Enterprise Throughout the Regions.

44

Why Space Elevators and Galactic Harbours?   This key question must be answered each time ISEC produces a book or report as we must encourage, enthrall, challenge, explain, and provide hope for readers. To anyone who looks up from their chair periodically and searches the heavens for the future of mankind, it is obvious that we are moving off-planet in a major fashion, and in the near future. Besides regular American, Russian and European space activities, the Chinese have landed a rover on the Moon and are planning a space station, the Indians have orbited a spacecraft around Mars, and the Japanese have a module attached to the International Space Station (ISS). The National Aeronautics and Space Administration (NASA) Jet Propulsion Lab (JPL) has identified over 1,300 near-Earth asteroids that are compatible with rapid trips made from Earth. There are three companies investing in mining resources on asteroids while there are multiple companies preparing to create small habitats on the Moon. In addition, there is a rocket company (SpaceX) that plans on building a colony of greater than 10,000 people on Mars within its CEO's lifetime.

To ensure that these dreams are encouraged and made successful, there must be a change in the approach to travel within our solar system. The cost to orbit must become a very small part of the overall investment and the arena must support infrastructures that can be used many times, not thrown away each time they are used. When one looks at the concept of space elevators, the answer is obvious. The future of humanity's travel within our solar system requires space elevators that provide access to space and that have the following strengths:

- Routine [daily],
- Revolutionarily inexpensive [<$100 per kg]
- Commercial development similar to bridge building
- Permanent infrastructure [24/7/365/50 years]
- Environmentally sound
- Safe and reliable [no shake, rattle and roll]
- Low risk lifting
- Low probability of creating orbital debris
- Redundant paths as multiple sets of space elevators become operational
- Massive loads per day [starts at 20 metric tons]
- Opens up tremendous design opportunities for users
- Optimized for geostationary orbit altitude and beyond

The bottom line for space elevators and the solar system is that they open up humanity's hopes and needs to expand beyond the limited resources and environment of our planet Earth. A space elevator is the enabling infrastructure ensuring humanity's growth towards the stars. There are two main reasons why the human race needs space elevator infrastructures:

- The realization that chemical rockets cannot get us to and beyond Low Earth Orbit (LEO) economically
- The recognition that the 'Space Option' may enable solutions to some of Earth's current limitations (energy, resources, removing nuclear waste etc.)

What kind of specific benefits could we expect to see from a functioning Space Elevator? As with the transcontinental railway, it's impossible to foretell all of the uses of such an infrastructure, but here are some possibilities.

- Large scale manufacturing in a zero-g environment. If corporations can build manufacturing facilities in space at an affordable price, they will do so. Right now, the cost and weight penalties are too prohibitive to even consider the idea. A space elevator would change that.
- Colonization of the moon, Mars and other planets and satellites. Currently, establishing and supplying a 6 or 8 person science station on the moon (let alone Mars or anywhere else) is probably at the very limit of our capabilities. Allowing hundreds (or even thousands) of tons to be launched into space every day would allow us to colonize these other worlds. This would provide an insurance policy for humanity, an outlet for those with a pioneering spirit and, almost certainly, increased benefits here on earth as commerce between our planet and others was established.
- Space Tourism – A Space Elevator could provide a way that most of us could visit space, and even stay for a while if we wanted to.
- Clean Power – Though there are many debates about the economics of establishing solar power satellites to provide earth with clean, limitless power, there is no doubt that to do so will require the capability to launch enormous quantities of materials into space. Only a Space Elevator can give us that capability.
- More and cheaper satellites. Satellite technology has provided all of us with enormous benefits, from DirecTV to weather satellites to increased national security. Being able to lower the cost and increase the reliability of satellite launches will lead to new technologies that right now we can't even imagine.

Scalable, inexpensive and reliable access to space will benefit all of us and a Space Elevator is the way to provide this capability.

How will the Space Elevator Work? Daily and routinely a Climber carrying cargo or people will be attached at the Earth Port. The Climbers will ascend the Tether, quickly leave the atmosphere and begin to make their way past Low Earth Orbit, between 160 and 2000 km up. While passing through this zone, cargo can be jettisoned to enter its own orbit around the earth. After four to five days, the Climber will reach Geosynchronous Orbit where more cargo will be detached. The cargo that remains on the tether above Geosynchronous Orbit will be moving faster than required to stay in orbit and can be detached and sent to destinations such as the Moon or Mars. The Climbers will then ascend to the end of the Tether where they will become part of the Apex Anchor as counter-weight. Several Climbers will be on the Tether at all times, each carrying their own small propulsion systems to 'move' the Tether out of the way of orbiting satellites and large space debris. Smaller space debris will be allowed to impact the Tether with the resulting damage taken care of by the Maintenance Climbers. Maintenance Climbers will be a constant companion of the Tether. They will travel the tether, continuously inspecting it and making repairs.

How does the Space Elevator stay vertical? Imagine you are holding a rope with a weight attached to the end. If you swing the rope in a circle at a sufficient speed, the rope will become taut, revolving about your hand. The force pulling the rope taut is known as centrifugal force. This same centrifugal force, generated by the rotation of the earth, will pull the Space Elevator Tether upwards into space (outwards from the earth).

Who invented the Space Elevator? The idea of a Space Elevator can be attributed to several different visionaries spread over more than one hundred years. In 1895 a Russian scientist named Konstantin Tsiolkovsky first proposed a tower into space. In 1959 another Russian scientist, Yuri Artsutanov came up with the idea of a tensile structure, something being pulled away rather than built up, to get into space. This idea used a satellite in Geosynchronous Orbit (GEO) to send a Tether down to the earth. In 1966 the idea moved in the U.S. with four American scientists writing an article about their "sky-hook" in the journal Science. American Jerome Pearson independently 'discovered' the idea of a Space Elevator and, in 1975 published his concept of the "Orbital Tower". By 1979 the concept was being spread to a larger audience by Arthur C. Clark in his novel The Fountains of Paradise. Today, the co-inventors of Artsutanov and Pearson are recognized as co-inventors of the concept with Dr, Edwards providing a solid design for the modern day [achievable] space elevator in 2002. see: Raitt, David, *Space Elevators: A History*, ISEC Report 2017.

How strong does the material have to be? The first important term for this question is Specific Strength. A spider web might not seem very strong but it has a high Specific Strength because of what it can hold versus its thickness. This is very important for a Space Elevator because all of the material will have to be lifted into space and because the Tether will have to be able to hold itself together over a great distance. The standard unit of measurement for Specific Strength is stress/density or Pascal/(kg/m3), for our purposes this can be adjusted to be GPa-cc/g (1Gpa-cc/g = 1 million Pascal/(kg/m3)). For simplicity ISEC has adopted the measurement scale of Yuri's, named after Yuri Artsutanov, where 1 MYuri is equal to 1 GPa-cc/g. Steel wire has a specific strength of about .5MYuri. Now we enter the realm of what is technically needed to build a Tether into space versus what is required to make a practical Space Elevator. A Tether with a specific strength of 25MYuri could be built but it would require a lot of mass and would not really be able to lift much. In the Space Elevator Feasibility Condition, the Spaceward Foundation's Ben Shelef discusses this problem in detail and shows how several factors enter into the question. The bottom line is that stronger is better with 30-40 MYuri's being the best bet for a practical Space Elevator, well within the predicted limits for carbon nanotubes and single crystal graphene. Less initial material and more payload to orbit will increase the rate at which a Space Elevator becomes a profitable venture. The recent discovery and production of a 0.5 x 0.1 meter single crystal of carbon atoms one layer thick has opened up the real possibility that a tether can be developed in the next few years.

How will space debris environment be handled? ISEC realizes that the density of space debris could become a serious hazard in the future. The 2011 ISEC Study Report presented an honest look at the space debris density numbers, where the Space Elevator is most vulnerable, and what can be done about the problem. It shows that space debris is a manageable problem, giving proper foresight and engineering. [note: the conclusions are still valid even with the increase in numbers in the last few years] The key is that ISEC believes the future will include a large global effort lowering the threats from space debris. ISEC depends strongly upon the future space community actively addressing this environment pollution problem with a positive approach before our first tethers in the 2030 time period. ISEC and the space community looking at the idea of a Space Elevator for the first time are concerned about how the ever-growing problem of Space Debris will affect it. We know the space elevator can safely operate in the environment; however, it would be beneficial if the global space community reduces the hazardous conditions.

Appendix B: Space Elevator Lexicon

After many meetings and discussions with players from around the world, a layout of a space elevator is shown below with a set of terms and their explanation following:

Apex Anchor Node & Region	LEO Gate	Earth Port and Region
Mars Gate	Lunar Gravity Center	- Earth Terminus
Moon Gate	Mars Gravity Center	- Floating Operations Platform
GEO Node and Region	Tether Climbers	Headquarters and Primary Operations Center

The following pages show the basics of the space elevator with definition of terms and figures that help define our concepts of space elevators. The following sections are expanded upon: SE Lexicon [with figure], Terminology Table, ISEC's Galactic Harbour Strategic Approach [with figure], SE Regions, SE Developmental Sequences, and our destinations.

Figure: Space Elevator System

Table of Suggested Terminology

Terminology	Explanation
Access City	Earth Port Access City will be the principle location where the majority of supplies/payloads depart from in route to the Earth Port. It should be the location for the HQ/POC and within 2,500 kilometers of Earth Port Region.
Apex Anchor	A complex of activity is located at the end of the Space Elevator providing counterweight stability for the space elevator as a large end mass. Attached at the end of the tether will be a complex of Apex Anchor elements such as; reel-in/reel-out capability, thrusters to maintain stability, command and control elements, etc.. [Note: nothing stays at that altitude unless attached to a tether]
Apex Anchor Region	The region around the Apex Anchor is defined by the amount of motion expected at the full extension of the tether. The region is the volume swept out by the end of the tether during normal operations. When two or more space elevators are operating together, the region spreads to the volume between.
Boron-Nitride Nanotube (BNNT)	High Tensile Strength material under development
Capability On Ramps leading to FOC	Time after IOC when new businesses / capabilities are added to system [7th sequence step]
Carbon Nanotube (CNT)	High Tensile Strength material under development
Climbers [Tether Climbers]	Vehicle able to climb or lower itself on the tether
Deployment	Releasing the tether from the GEO construction up and or down during the initial phase of construction
Earth Anchor (Tether Terminus)	Earth Terminus for space elevator
Earth Port	A complex located at the Earth terminus of the tether to support its functions. These mission elements are spread out within the Earth Port Region. When there are two or more termini of tethers, the Earth Port reaches across the region and is considered one Earth Port.
Earth Port Region	The volumetric region around each Earth Port to include a space elevator column for each tether and the space between multiple tethers when they operate together. The Earth Port Region will include the vertical volume through the atmosphere up to where the space elevator tether climbers start operations in the vacuum and down to the ocean floor.
Floating Operations Platform	The Op's Center for the activities at the Earth Port or Earth Terminus
Full (Final) Operational Capability	Design for full capability of the space elevator [8th sequence step]
GEO Node	The complex of Space Elevator activities positioned in the Space Elevator GEO Region of the Geosynchronous belt [36,000 kms altitude]; directly above the Earth Port. There will be several sub nodes; one for each tether, one for a central main operating platform, one for each "parking lot", and others.
GEO Region	Encompasses all volume swept out by the tether around the Geosynchronous altitude, as well as the orbits of the various support and service spacecraft "assigned" to the GEO Region. When two or more space elevators are operating together, the region includes each and the volume between elevators.
Headquarters and Primary Operations Center [HQ/POC]	Location for the Operations and Business Centers – probably other than at Earth Port – more likely near Space Elevator Access City
Initial Operational Capability	A term to describe the time when the space elevator is prepared to operate for commercial profit – robotically [6th sequence step]
International Academy of Astronautics (IAA)	International Association focusing upon space capabilities with approximately 1,000 elected members.
International Space Elevator Consortium (ISEC)	Association whose vision is: A world with inexpensive, safe, routine, and efficient access to space for the benefit of all mankind.
Japanese Space Elevator Association	JSEA handles all the space elevator activities for universities and STEM activities. Also handles the global aspects of space elevators.
Japanese Space Agency (JAXA)	Japanese government organization responsible for space systems and space operations.
Length Overall	Full length of the space elevator, est. from 96,000 to 100,000 km
LEO Gate	Elliptical release point for LEO – roughly 24,000 kms altitude
Limited Operational Capability	Early utilization of a "starter" tether in parallel with testing and further development [5th sequence step]
Lunar Gate (Moon Gate)	Release Point towards Moon – roughly 47,000 kms altitude
Lunar Gravity Center	Point on Tether with Lunar gravity similarity – 8,900 kms altitude
Marine Node (Earth Port)	Earth Terminus for space elevator
Mars Gate	Release Point to Mars – roughly 57,000 kms altitude
Mars Gravity Center	Point on Tether with Mars gravity similarity – 3,900 kms altitude
Ocean Going Vehicle (OGV)	Vehicle able to travel over the open ocean
Operational Testing	Key developmental phase when checking out capability [4th sequence step]
Pathfinder	In-orbit testing of space elevator with as many segments represented as possible [1st sequence step]
Primary Operations Center (POC)	Center of all activities for the space elevator. Could be distributed or centralized.

Seed Tether [Ribbon]	The initial tether lowered from GEO altitude which would then be built up to become the space elevator tether [2nd sequence step]
Single String Testing	Single string tests are tests conducted of a selected set of Space Elevator functions; aligned and operating. In early forms, single string testing could be an end-to-end simulation of a segment. Later, hardware is inserted in the string to add realism. Testing the initial tether after deployment would be a key single string test. [3rd sequence step]
Space Elevator Column	The volume swept out during normal operations starting at the Earth Port [a circular area within which it operates] and extending through the GEO Region up to the Apex Region. This column of space will be monitored, restricted, and coordinated with all who wish to transverse the volume.
Tether	100,000 km long woven ribbon of space elevator with sufficient strength to weight ratio to enable an elevator [CNT material probably]
Tether Climbers	Vehicle able to climb or lower itself on the tether, as well as releasing or capturing satellites for transportation or orbital insertion.

ISEC's Galactic Harbour Strategic Approach

One of the principle elements of the International Space Elevator Consortium's (ISEC) action plan towards an operational space elevator is to understand its customer utilization. To fully understand the potential application for commercial ventures on the space elevator, the concept of a Galactic Harbour surfaced. Galactic Harbour represents continuous operations moving customer payloads on multiple space elevators from entry ports to exit ports. These locations would most logically be an Earth Port where customers have their payloads loaded onto space elevators and their release points at multiple altitudes per the desires of the customers. The Galactic Harbour would then be the volume incorporating multiple Earth Ports [on the ocean, with incoming and outgoing ships/helicopters and airplanes] and then stretch up in a cylindrical shape to include tethers and other aspects out to the Apex Anchors.

Galactic Harbour as the Unification of Transportation and Enterprise

Space Elevator Transportation is the "main channel" in the Galactic Harbour	Businesses flourish as a part of the Space Elevator Enterprise System
GEO Node	Business support to Operational Satellites
Earth Port	Power and products delivered to Earth
Apex Region	Interplanetary Efforts within reach
Tether Climbers	Research
Tether System	
HQ/POC	

Our Destinations

- *The Initial Operational Capability (IOC) consists of a system comprised of two space elevators with one Earth Port and two terrestrial terminus, two Apex Anchors each with 100,000 km tethers, multiple tether climbers and a single Headquarters and Primary Operations Center. This system will be capable of moving significant payload tonnage [20 Metric ton] to GEO and beyond several times a week from each space elevator.*

- *The Full Operational Capability (FOC) contains two tethers per elevator system (100,000 km strong tether), each with a tether terminus platform inside the Earth Port, GEO Node, Apex Anchor, and with a single Headquarters and Primary Operations Center. This system will be capable of moving an estimated 70 Metric tons to GEO and beyond several times a week (with passengers).*

Space Elevator Regions

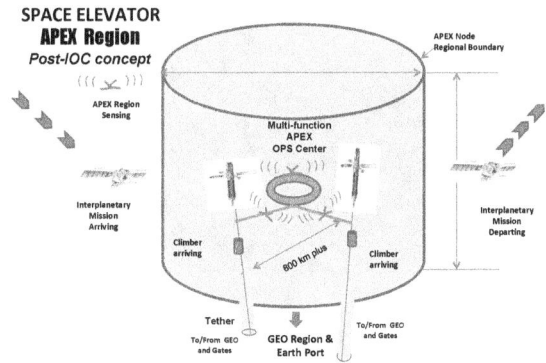

Apex Anchor Region: The region around the Apex Anchor is defined by the amount of motion expected at the full extension of the tether. The region is the volume swept out by the end of the tether during normal operations. When two or more space elevators are operating together, the region spreads to the volume between.

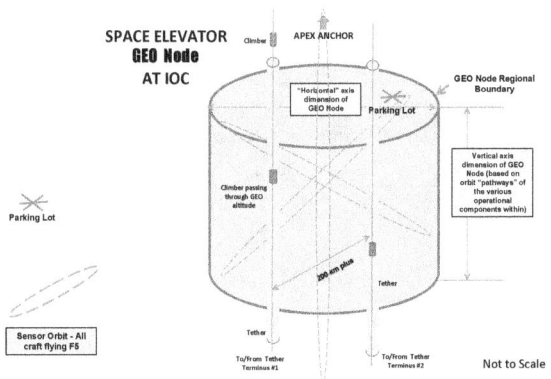

GEO Region: Encompasses all volume swept out by the tether around the Geosynchronous altitude, as well as the orbits of the various support and service spacecraft "assigned" to the GEO Region. When two or more space elevators are operating together, the region includes each and the volume between elevators.

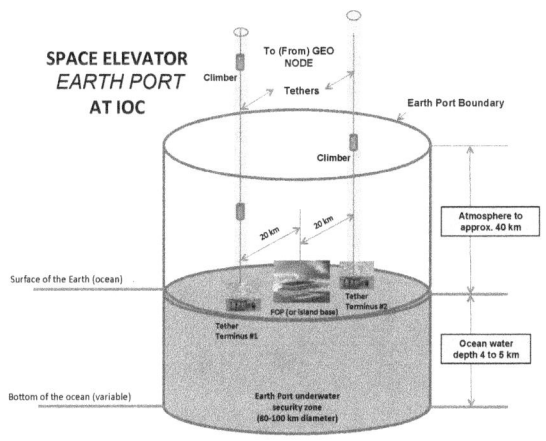

Earth Port Region: The volumetric region around each Earth Port to include a space elevator column for each tether and the space between multiple tethers when they operate together. The Earth Port Region will include the vertical volume through the atmosphere up to where the space elevator tether climbers start operations in the vacuum and down to the ocean floor.

Space Elevator Development Sequence

Setting the stage with a typical life cycle phase schedule for developing space systems.
Space Elevators are still, very much, in the Concept Development phase.

Typical Project Life Cycle Phases

Project Life Cycle Phases	Pre Phase A: Concept Study	Phase A: Concept & Technology Development	Phase B: Preliminary Design and Technology Completion	Phase C: Final Design & Fabrication	Phase D: System Assembly, Integration & Test, Launch	Phase E: Operations & Sustainment	Phase F: Closeout
Reviews -Mission		MCR	MDR				
Reviews -System		SRR SDR	PDR	CDR	ORR FRR		

Formulation Phase
(More Academic level efforts are required)

We are still here.

Implementation Phase (Space Agency, Private sector, Industries, etc.)

Space Elevator Development

Space Elevator On-orbit Assembly, Checkout, and Operations

<Notes>
MCR: Mission Concept Review, MDR: Mission Definition Review, SRR: System Requirements Review,
SDR: System Definition Review, PDR: Preliminary Design Review, CDR: Critical Design Review,
ORR: Operational Readiness Review, FRR: Flight Readiness Review
(Ref: NPR7123.1A NASA Systems Engineering Processes and Requirements w/Change 1 (11/04/09))

Special Sequence for Development of Initial Space Elevators

9. Pathfinder
10. Seed Tether,
11. Single String Testing
12. Operational Testing,
13. Limited Operational Capability (LOC),
14. Initial Operational Capability (IOC),
15. Capability On Ramps leading to FOC
16. Full Operational Capability (FOC)

Note: it is critical to initiate second tether for backup as soon as practical – in this sequence, Deployment 2 should be right After Single String Testing, with IOC only when two space elevators are up.

Appendix C: Summary ISEC Studies

Design Considerations for a Multi-Stage Space Elevator [2019]

To build a space elevator, the toughest challenge is to find material that is strong enough for a self-supporting tether. Building it in multiple stages is a way of overcoming that challenge. Using the concept of dynamically supported structures, it is possible to build upwards from the earth's surface and provide supports for the lowest parts of the tether, where gravity is strongest. A five-stage design would support a tether made of carbon fiber yarn that is commercially available today. A two-stage design can support a tether with less than one-third of the strength previously thought necessary. The study report analyses the proposal in detail, covering the underlying physics and technology, design options and prototyping work. Authors: John M. Knapman, Peter Glaskowsky, Dan Gleeson, Vern Hall, Dennis Wright, Michael Fitzgerald, Peter Swan

Design Considerations for the
Multi-stage Space Elevator

John M. Knapman
Peter Glaskowsky
Dan Gleeson
Vern Hall
Dennis Wright
Michael Fitzgerald
Peter A. Swan

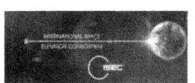

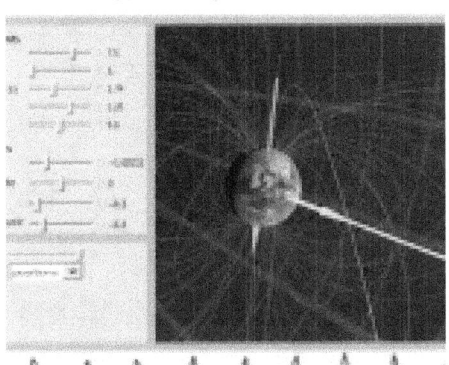

Design Considerations for a Software Space Elevator Simulator [2018]

This study report gives a detailed analysis of all the design considerations for a Software Space Elevator Simulator. From the Executive Summary: As with all large, modern engineering projects, detailed computer simulations of the space elevator will be essential during its design, construction and operational phases. Within the context of these phases, this study enumerated 14 use cases which the simulation software must address, ranging from 3D dynamics and electrodynamics calculations of space elevator motion, to the effects of payload capture and release at various points along the tether, to the effects of friction arising from the interaction of the space elevator climber with the tether. Proceeding from these use cases, requirements were imposed on the software design and an outline for its development was sketched. Authors: Dennis H. Wright, Steven Avery, John Knapman, Martin Lades, Paul Roubekas, Pete A. Swan

Design Considerations of a Space Elevator Apex Anchor and GEO Node [2017]

This year, ISEC chose to address the design considerations for the Apex Anchor and Geosynchronous Earth Orbit (GEO) Node. As was discussed in the Architectures and Roadmap

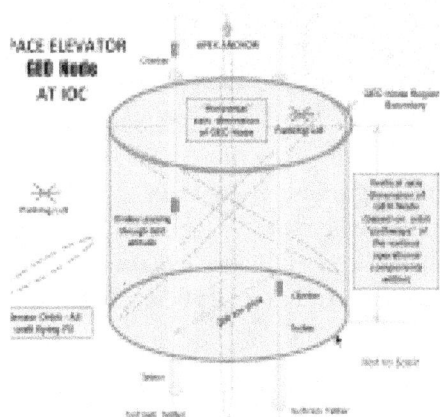

report, ISEC understands where the technologies are today and where we would like them to be in order to reach Initial Operational Capability (IOC). The goal of this study team is to add to the "body of knowledge" relative to the two topics addressed herein. To ensure complete understanding during this study report, the following definitions were developed: Space Elevator Column, Earth Port & Earth Port Region, GEO Node & GEO Region, and Apex Anchor & Apex Anchor Region. In addition, the needs [functional requirements] were discussed for each of those regions and complexes. Throughout the text, the initial destination is described as the IOC for the Space Transportation System. The Space Elevator Transportation System is comprised of one Earth Port with two tether termini, multiple Apex Anchors supporting 100,000 km Tethers, 14 Tether Climbers, and a single Headquarters and Primary Operations Center. The GEO Node supports the Space Elevator Transportation System with a range of "overhead' functions; e. g. test, safety, and support. Authors: Michael Fitzgerald, Vern Hall, Peter Swan, and Cathy Swan.

Design Considerations of a Space Elevator Earthport [2016]

This study report provides the International Space Elevator Consortium's (ISEC) view of the Earth Port (formerly known as the Marine Node) of a Space Elevator system. The

Earth Port: Serves as a mechanical and dynamical termination of the space elevator tether; Serves as a port for receiving and sending Ocean Going Vessels (OGVs); Provides landing pads for helicopters from the OGVs; Serves as a facility for attaching and detaching payloads to and from tether climbers and attaching and detaching climbers to and from the tether; Provides tether climber power for the 40 km above the Floating Operations Platform (FOP); and, Provides food and accommodation for crew members as well as power, desalinization, waste management and other such support.

Authors: Robert E. 'Skip' Penny, Jr, Vern Hall, Peter Glaskowsky, and Sandee Schaeffer.

Space Elevator Survivability, Space Debris Mitigation[2011]

This report focuses on the issue of Space Debris in relation to a Space Elevator. Many people looking at the idea of a Space Elevator for the first time are concerned about how the ever-growing problem of Space Debris will affect it. This report gives an honest look at the numbers, where the Space Elevator is most vulnerable and what can be done about the problem. It shows that space debris is a manageable problem, giving proper foresight and engineering. Authors: Dr. Peter Swan, Cathy Swan and Robert "Skip" Penny.

Space Elevator Concept of Operations[2013]

This report describes and discusses a plausible Operations scenario for a Space Elevator. This report addresses initial commercial operations of a space elevator pair with robotic climbers. This report has been developed to help define a starting point for an initial space elevator infrastructure. It is assumed that there are two space elevators in place to ensure continuation of our escape from the gravity well. It also assumes that a sufficient number of climbers are available for delivering of spacecraft and other payloads to orbit, and, if required, return them to earth. In addition, this report is designed to be the initial operations concept from which many improvements will occur as future knowledge and experience drives infrastructure concept revisions. Authors: Dr. Peter Swan, Cathy Swan and Robert "Skip" Penny.

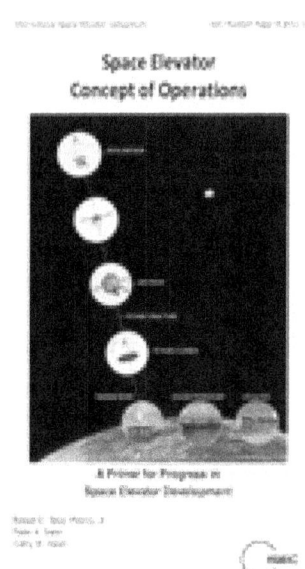

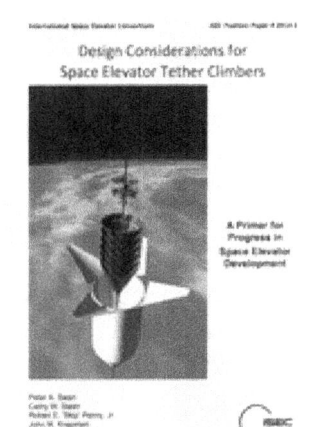

Design Considerations for Space Elevator Tether Climbers[2014]

The subject selected for this 2013 study is the Space Elevator Tether Climber. The objective of the one year study was to survey current concepts and technologies related to tether climbers, identify critical issues, questions, and concerns, assess their impact on the development of space elevators, and project towards the future. Authors: Dr. Peter Swan, Cathy Swan, Robert "Skip" Penny, John Knapman and Peter Glaskowsky.

Space Elevator Architecture and Roadmaps –

This 2014 study report establishes a baseline roadmap for designing space elevators for the future. This study addresses critical aspects of space elevator infrastructures: basic architectures and how we will get there with a roadmap. The roadmaps will leverage desired paths to lower risks and identify approaches for pulling together the diverse concepts. The three architectures in the literature today are solid looks at various approaches, while not providing that key element of "how will we get there?" Each path from today to the successful implementation of a space elevator infrastructure must be identified and discussed with respect to hurdles and milestones.
Authors: Michael "Fitzer" Fitzgerald, Peter Swan, Cathy Swan and Robert "Skip" Penny.
Publication date: April, 2015

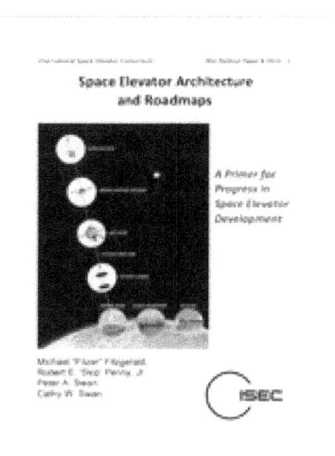

Appendix D: Summary of IAA Studies

After the creation of the International Space Elevator Consortium in 2008, the involvement with the global space community became a priority. One straightforward approach was to actively support academic studies dealing with the space elevator. The International Academy of Astronautics has conducted two studies, 2015 and 2019 publishing dates. https://iaaweb.org   A portion of their executive summaries will be shown.

"Space Elevators: An Assessment of the Technological Feasibility and the Way Forward." 2015

What are the questions for this study report?   This report addresses the simple and complex issues that have been identified through the development of space elevator concepts over the last decade. The report begins with a summary of those ideas in Edwards' and Westling's book "The Space Elevator" (2003). Out of these beginnings has risen a worldwide cadre focused upon their areas of expertise as applied to space elevator development and operational infrastructure. The report answers some basic questions about the feasibility of a space elevator infrastructure. A preview of the main questions and answers shows the depth and breadth of this Cosmic Study.

What is a space elevator?   A space elevator is a system for lifting payloads, and eventually people, from the Earth's surface into space. The one under consideration in this report consists of a tether 100,000km long balanced about a node in geosynchronous orbit (GEO) and reaching down to an anchor point on Earth. Electrically powered spacecraft, called tether climbers, travel up or down the tether at far lower costs [currently projected at $500/kg] than using rockets. In addition, the service the space elevator provides is a cargo capacity/throughput of two orders of magnitude larger than present rockets, with tremendously kinder environmental effects, and a miniscule potential for future space debris. Tether climbers can continue to the apex anchor – the point at 100,000km altitude – where their speed is sufficient for direct interplanetary travel.

Why a space elevator? The value and benefit of developing a space elevator infrastructure is even greater than earlier estimates, as it will change our approach to operations in space. Low cost, safe, reliable and flexible delivery of payloads to Geosynchronous Earth Orbit (GEO) and beyond could create an "off-planet" environment filled with opportunities ranging from commercial space systems to exploration of the solar system. Daily initiation of 20 metric ton climbers, safe delivery to GEO and beyond, and a projected price of $500 per kg, will open up the solar system and lead to many new commercial ventures. In addition, the radical change from chemical rockets and the low risk approach of climbing vertically at reasonable speeds will greatly reduce two major hazards that are dominant today: 1) the environmentally friendly, electrically driven, motors will have almost no hazardous material polluting the atmosphere, and 2) this delivery technique does not create orbital debris, especially in Low Earth Orbit (LEO).

Another major benefit will be in supporting human exploration. The first ten years will enable massive movement of equipment to GEO and beyond. Human exploration can leverage this tremendous capability by assembling large spacecraft at GEO with massive fuel loads delivered at $500/kg. After ten years of operations, humans should also be riding to GEO.

The benefits for humanity on Earth can be phenomenal. The ability to inexpensively deliver large quantities to orbit will enable capabilities stimulating an Earth renaissance. The facility to provide power to any location on the surface [space solar power satellites] will enable

development across the world. Several examples are that Africa could skip the last century of wires while the outback of countries like India or China would not have to burn coal and the Amazon region could retain more of its rain forests. In addition, the increase in communications and Earth resource satellites will remake the emergency warning systems of the world. Some intractable problems on the Earth's surface would also have solutions, such as the safe and secure delivery – and thus disposal - of nuclear waste to solar orbit.

Can it be done? The authors recognize that the whole project, especially the projected price per kilo, is dependent upon a strong, lightweight material that will enable the space elevator tether. The principal issue is material produceability at the strength, length and perfection needed to enable a 100,000km long tether. Almost all other issues surrounding each of the major segments have either been resolved in space before or are close to being space ready today. Only the tether material is at a high technological risk at this time. Chapter 3 goes into projections of material growth and increase in capabilities showing their potential with a good prospect of suitable material becoming available by the 2020s.

How would all the elements fit together to create a system of systems? Each of the early chapters addresses one of the major elements of space elevator infrastructure. As the study progresses, the reader moves from tether material to individual segments to systems level analyses. This sequence illustrates the parts of a space elevator infrastructure and then shows the operational view as it all fits together. In addition, in the market and financial chapters, the development of future space markets are projected with their funding profiles for the next 40 years.

What are the technical feasibilities of major space elevator elements? Each of the individual chapters describes major segments of the space elevator and discusses NASA Technical Readiness Levels and Risk Management trades to ensure the technical feasibilities can be assessed. The space elevator roadmaps show the approach from the current year [2013] to operational time periods. A factor for the future of space elevator infrastructure is the majority of components, subsystems, and segments have been developed before as components of other space systems [except for the tether material]. This leverage of 50 years' experience is invaluable and will enable development of space elevator segments in a timely manner.

The conclusions from this study fall into a few distinct categories:
- **Legal:** The space elevator can be accomplished within today's arena
- **Technology:** Its inherent strengths will improve the environment and reduce space debris in LEO and beyond. It can be accomplished with today's projection of where materials science and solar array efficiencies are headed. The critical capability improvement is in the space elevator tether materials, currently projected to achieve the necessary strength to weight ratio in the next 20 years. The space elevator will open up human spaceflight and decrease space debris and environmental impacts.
- **Business:** This mega-project will be successful for investors with a positive return on investment within 10 years after erection is complete.
- **Cultural:** This project will drive a renaissance on the surface of the Earth with its solutions to key problems, stimulation of travel throughout the solar system, with inexpensive and routine access to GEO and beyond.

Cosmic Study Result The authors have come to believe that the operation of a space elevator infrastructure will lead to a "game changing" experience in the space world. Each of the authors considers that the space elevator can be developed when the tether material is mature enough. Our final assessments are:

A Space Elevator appears feasible, with the realization that risks
must be mitigated through technological progress. and,
A Space Elevator infrastructure will be achievable through a major global
enterprise.

"Road to Space Elevator Era." 2019

This study report summarizes the assessment of the space elevator as of the summer of 2018. The encouraging aspect is that the space elevator community has been reinvigorated and is pulling together experiments and test programs to push the technology along the path to readiness. Several of these break through are the ones we were searching for after completion of the first IAA study. We see the way forward! The global needs for a space elevator are remarkable. When the price to geosynchronous orbit is lowered to one hundredth of the price of launching by rocket, the whole situation changes as to access to orbit. However, the real strengths are not only price but massive movement and other characteristics such as routine, daily, safe and no shake-rattle-roll of launch. The environmentally friendly lifts will be an important aspect of implementing

space elevators vs. rockets in the long run. One key to recognize is that we move from individual events to continuous operations of an infrastructure with the space elevator. We would move to a system with the costs representing recurring expenses, not replacement costs. The concept is to move to a "bridge to space," not a system of individual rocket launches. The question on the table is "are we actually on the road to a space elevator?" The study answers that question in a positive manner. Yes, we are on the road to the space elevator era!

This study report assesses the global needs for a space elevator and then lays out functional requirements that can lead to technological needs and identification of processes for development. The development of the needs and requirements lead to the chapters assessing the critical technologies and then recommends the risk reduction approaches for segment and

Road to Space Elevator Era

Editors:
Peter A. Swan
David I. Raitt
John M. Knapman
Akira Tsuchida
Michael Fitzgerald

International Academy of Astronautics

system level verification. In addition, Chapter Six defines validation testing for the customer, as a milestone towards total project funding.

As the goals of the study were approached systematically during the study, the results were presented in the form of conclusions and recommendations. When one looks at all the various technologies and where they are in the technology readiness level (TRL) evaluations common to NASA projects, the team has the following conclusions:

- The Earth Port is buildable with today's available technologies and engineering expertise.
- The Headquarters and Operations Centers are buildable today.
- The tether climber is so similar to a normal satellite design of today that there is

no real technological or engineering challenge; except the interface with the tether material. As there is a lack of information of the chosen material for the tether, some engineering must be resolved at a later time.

- The GEO Node and GEO Region technologies are understandable and not an issue during development.

- The Apex Anchor will be a challenge as its role is key to the building of the space elevator, but not an engineering and technological issue.

- However, the tether material is the pacing item for the development of the space elevator. Currently, there are three viable materials that could grow into the needed strong-enough and long-enough material for a space elevator: carbon nanotubes, boron nitride nanotubes, and continuous growth graphene. The community waits for those materials to mature to the level that can be implemented into a space elevator tether 100,000km long and strong enough to support its own weight plus multiple tether climbers against the pull of gravity. At 100 million metres long, a space elevator tether is a major engineering challenge. Recent investigations explored the possibility for making single crystal graphene by a continuous process using liquid metal. Making this a viable practical manufacturing process will be a significant effort over a period of years and probably many millions of dollars. However, such a process would create graphene products for many multi-billion dollar markets on the way to making the tether material. For this reason there is a credible return on investment case for manufacturing the material in practise. This means it is highly possible that continuous single crystal graphene will be manufactured in the coming years and this material should be considered in any forward thinking about a space elevator tether.

- Learning to extract best performance reel-out of imperfect materials is a common engineering problem; rare indeed is the design where all constraints and criteria are fully satisfied by a single solution. Two challenges that separate us from a tether become clear. The first, a challenge of assembly—how do we ensure uniform load distribution in our material, so that we can bring the nanoscale properties up to our macroscale application. The second, a challenge of production—how to scale existing processes up to produce the volume needed. Neither of these challenges requires fundamentally new science or engineering. They require continued application of existing knowledge and skills.

Based on these conclusions, a number of recommendations can be made - the primary one of which is to encourage and support specific strength material development with the purpose of making them long enough and strong enough for space elevator tethers. The essence of the whole study report is that a broad group of space professionals gathered together and assessed the status of the space elevator development. Each of them contributed their expertise and then came to similar conclusions about the space elevator progress. It should be noted that, even though many of the references have several authors, for convenience in the text only the first named author is given.

Space Elevators seem feasible [reinforces IAA's 2013 study conclusion],
and
Space Elevator development initiation is nearer than most think!

60

Appendix E: Summary of ISEC Architectural Notes

Space Elevator Architecture Note #1
Space Elevator Enterprise will be based on a Modular Construction

The Space Elevator Enterprise will be based on a Modular Construction, Modular Growth and Modular Operating Standard, such as;
     Building the Space Elevator -- Modular Construction
     Growing the Space Elevator -- Modular Growth
     Operating the Space Elevator -- Modular Operations

Space Elevator Architecture Note #2
Space Elevator Enterprise will encounter robust & diverse on-orbit business

When it achieves IOC, the Space Elevator transportation system will encounter robust & diverse on-orbit business activity.
     Future Business in Space -- Now
     Future Business in Space – Some future visions, as seen now.
     The business offerings of the Space Elevator

Space Elevator Architecture Note #3
Space Elevator Enterprise will need a Business Capture Plan

We need a Business Capture Plan and a working group that engages with on-orbit business activities; joining in, if and when we can.
     Space Elevator as a business – We aren't trying to be one, yet
     Space Elevator business enterprise – As we see it now
     Space Elevator business enterprise planning – the timing of this

Space Elevator Architecture Note #4
Space Elevator Enterprise will be based on a Modular Construction

We need a Baseline. The Baseline of the Space Elevator is still changing and that is a good thing; but only for now.
     Space Elevator baseline – We don't have one and we might need two:
     Space Elevator Transportation system – Baseline #1
     Space Elevator Business Enterprise – Baseline #2
     Space Elevator Business Enterprise – Baseline relationships

Space Elevator Architecture Note #5
Emergence of the Galactic Harbour Vision
The emergence of The Galactic Harbour Vision: ISEC sees a great Space Elevator Transportation system that opens Space; in the same way the railroad opened up the American west. It is a marvel of Kepler's physics; and a marvel of humankind. We see the Galactic Harbour importing the needed (Power and unique minerals); exporting the noxious (Nuclear waste and Space debris); and supporting interplanetary exploration.

Space Elevator Architecture Note #6
Space Elevator Developmental Sequences (1&2)

The Space Elevator "Sequences" --- A discussion of ISEC's basic technology and engineering development approach.
     First, start the design process for the Space Elevator Transportation system
     Sequences #1 – Pathfinder experiments

Sequences #2 – Seed Tether / Seed Events
Sequences have sequences

Space Elevator Architecture Note #7
Space Elevator Developmental Sequences (3, 4, & 5)

The discussion about the "Sequences" program continues. The need for an extensive on-orbit testing program emerges.
    Sequences #3 – Single String Testing
    Sequences #4 – Operational Testing
    Sequences #5 – Limited Operational Capability
    Repetition is not a bad thing - Repetition is not a bad thing

Space Elevator Architecture Note #8
Space Elevator Developmental Sequences (6, 7, & 8)

The culminating steps in the "Sequences" program are discussed
    Sequences #6 – Initial Operational Capability (IOC)
    Sequences #7 – Capability On Ramps
    Sequences #8 – Full Operational Capability (FOC)

Space Elevator Architecture Note #9
Space Elevator Strategic Approach
The Strategy to build the Space Elevator : This Note introduces ISEC's "Strategic Approach" for the development of the Space Elevator. It is a discussion of how to turn a long-term vision into a long-term "plan".

Space Elevator Architecture Note #10
Space Elevator Planning Horizon is 30 Years
Our Long-Term (30+ year) Planning Horizon may be the hurdle to possible venture funding from industry: This note discusses how we can seek support from industry funding sources. These sources provide funding grants for development of new technologies and new concepts. ISEC must convince them that our approaches will provide predictable return on their investments.

Space Elevator Architecture Note #11
Planning Horizon is Different than Having a Plan
We have a Planning Horizon reaching out to 30 years into the future. That is different than having a plan for the next 30 years. The key word is horizon. Not "over the horizon" like a fancy radar or a battlefield cavalry scout. Not "short of the horizon", like something that is well within our view. ISEC expects our Space Elevator to be valuable for a while after IOC; out to the horizon

Space Elevator Architecture Note #12
Space Elevator Conference (2017) is Highlighted by Progress
The International Space Elevator Consortium met in Seattle for its 2017 Conference. Boeing's Museum of Flight sets the right tone for our musings; and, we all had a great time. ISEC and Conference attendees embraced a new vision of its Space Elevator; the Galactic Harbour. It will be a transportation force in the future, an enabler of robust space-based enterprise, and the initial infrastructure of the 3rd dimension of Earth's transportation and logistics system.

Space Elevator Architecture Note #13
Communications within the Galactic Harbour
This Note discusses how can we manage all these space objects with the entirety of the Galactic Harbour. Safety and efficiency calls for communications with all the objects in our inventory, keeping all of them under positive control; even though the objects are spread over a few zillion cubic kilometers of the great unknown.

Space Elevator Architecture Note #14
Galactic Harbour Delineation
The concept of the Space Elevator Transportation System versus the Space Elevator Enterprise System is discussed. There must be expansive discussions of the essential functions of the Space Elevator Transportation System and – then- of the Space Elevator Enterprise System. For the sake of order and simplicity, all currently identified functions need to be placed in one of four stacks; 1) Transportation, 2) Enterprise, 3) both (or shared), or 4) neither (customer provided).

Space Elevator Architecture Note #15
The Galactic Harbour's Full Operational Capability
First, let us recall our definition of FOC in terms of a development phase. The Full Operational Capability (FOC) is achieved after Initial Operational Capability (IOC) is achieved AND after adding some functions to the Space Elevator via Step 7 in the Sequences; the On-Ramp Step. For those new to the game, "Sequences" are the 8 steps we intend to go through to attain an operational Space Elevator (see notes 6, 7, and 8).

Space Elevator Architecture Note #16
Space Elevator Transportation System, Ready?
As a preview, ISEC's preliminary Technology and Engineering Readiness is summarized in the note. When one looks at all the various technologies and where they are in the Technology Readiness Level evaluations common to NASA projects, ISEC sees these preliminary assessments as completed and ready to move into segment and system level testing by industry.

Space Elevator Architecture Note #17
Galactic Harbour Strategic Approach must become our Plan
Our strategy is to link the Space Elevator Transportation System to the Space Elevator Enterprise; within a Unifying Vision known as the Galactic Harbour. The Strategic Approach is ISEC's guiding theme for the technical development of a Space Elevator. The Space Elevator Transportation System will be the core, priority construction activity; and, its success will be the foundation of the Space Elevator Enterprise System. They will be built in a manner separate from each other but not in isolation. This "separate but not segregated" paradigm establishes both the prioritization and collaboration between and within our near parallel development efforts.

Space Elevator Architecture Note #18
ISEC Must begin a Broad Outreach Program
There are any number issues that could affect us at the destination our progress. We need to "check the schedule"; and see how a Space Elevator fits into the future. That is, the future as seen by others. We need to reach out to a variety of legal, diplomatic, technology, and jurisdictional authorities; and see how they see the Space Elevator fitting

into their future. If need be, ISEC should offer an enlightened view … so they see things our way. It is our teaching moments and we need to make some plans. We need to form our story (stories) and stick to it (them).

Space Elevator Architecture Note #19

Space Elevator Needs to Present out Story

ISEC is having a small conference in Seattle in Aug 2018. Our theme is that the Space Elevator Transportation System development is closer than you think. That is our story. ISEC is going to review the seven positive statements of the preliminary Technology Readiness Assessment and make sure that development is ready for each of the segments of the Transportation System. ISEC is going to do that self-examination with thoroughness. That is our story and we are sticking to it.

Space Elevator Architecture Note #20

Architecture Engineering Baseline -- Change Management

We should be prepared for a decade of changes; followed by another series of changes; and after that, more changes. So, we need an orderly change approach. An immutable change approach is essential so that all working on the Space Elevator are working on the same thing. The "change approach" matures into a detailed change process; used during the Elevator's design and development. Our modular design approach (See Architecture Note #1) and our application of sequenced Architecture Engineering principals (See Architecture Notes #6 through #8) must adhere to the fundamental Architecture theme that mission performance in one segment can affect mission performance in other segments. Thus, we foresee a change approach which is used between now and the end of Validation demonstrations (see the Architectures and Roadmap Report); and a change process (used during design and development).

Space Elevator Architecture Note #22

Robots and the Space Elevator

The history of space launch and operations, robotically, in retrospect, was remarkable. But - it is noted here – from those critical early moments at launch until mission success and de-orbit it was operated remotely and robotically. So will it be with the Space Elevator. Humans will slide the payload into the Climber, conduct final safety checks: and off it will go -- operated remotely and robotically. What would you expect? The Space Elevator is the basis for safe, reliable, routine, environmentally clean access to space. It is a job made for robots.

Space Elevator Architecture Note #23

Do you know the way to Anywhere?

One of the major benefits of a Space Elevator is the energy that can be used by just being released from the Apex. The Apex is moving along at a rate based on the spinning Earth - to which it is attached. Calculations show that velocity is nearly 8 kilometers per second. That speed is the galactic free ride to anywhere; except you need to know the way. As long as we are getting help with planes, planar crossings, coplanar approaches, orbits, departure times, vectors, and velocities to approach our destination; we might as well ask for a list of destinations. Let's see - the Moon, Mars, several asteroids, moons around Mars, and probably more destinations. Sort of like an interplanetary bus schedule, with different information columns; velocities needed, departure times, plane changes, duration of trips, and so on. Our departure from the Apex needs to be effectively aimed. The highways in Space are the orbits of the heavenly bodies.

Space Elevator Architecture Note #24

The Path to Tech Readiness!

In the last 6 years, ISEC's Technology Maturation approach has melded with a better definition of the Space Elevator engineering solution. The 2014 publication of ISEC's "Architecture and Roadmap" Report removed the shroud of mystery and myth from the Elevator's scope and complexity. In the last year, the International Space Elevator Consortium advocated that the basic technologies needs were available, and each segment of the Space Elevator Transportation System was ready for engineering validation. Phase two will begin soon after phase one completion. Industry involvement is an imperative. Phase two activities are driven by six major activities:
Examine Industry's production foundation
Determine if the segments can be built
Assess schedule & technical risk:
Delineate design criteria
Set criteria and standards to enter the Design Validation Phase
Baseline operations performance:

Space Elevator Architecture Note #25

Space Elevator Architecture's Debris Mitigation Roles

ISEC believes that debris mitigation concepts will be built, operating, and thriving before the Space Elevator Transportation System reaches operational status. To that end, this paper serves as the initial characterization of how the Elevator can allocate the needed performance to a system then available. there are five parts to the ISEC mitigation approach: Debris Alerts available, Debris Size identified, Tether movement upon demand, Sentry Concept developed, and Recovery approaches included.

Space Elevator Architecture Note #26

Road Signs or not, the Space Elevator Transportation System changes everything!

The Space Elevator Transportation System will make a dramatic difference on intergalactic travel. It changes everything about travel within our solar system. Travelers and supplies should not have to wait for many months for the proper alignment of orbit planes to get to Mars. Departures can be every day! The duration of travel will also be much less than the long trip we have heard. Our vision is simple è Free, Fast and Daily Delivery! No waiting! Package up your stuff, send it up the Elevator, and depart according to the bus schedule.

Space Elevator Architecture Note #27

Space Elevator Mission Support Equals Mission Success

Begin a collaborative mission definition & documentation process of the missions to Mars, the Moon and elsewhere. The primary premise is that interplanetary missions will be fully and completely supported by a Space Elevator Transportation System; almost concurrently with the exploratory manned mission elements at the outset; that outpost reinforced by the Space Elevator delivered supplies and logistics needed to sustain and grow.

Appendix F: List of Space Elevator References

List of Space Elevator References

Dear Space Elevator Researcher – A summary of papers and books about space elevators is shown below with a breakout as shown in the table of categories. Most are on the web already, but if you have trouble finding an article, [or wish to update the list] please email us and we will try to connect you with the author – **inbox@isec.org**. *[as of Sept 2019]*

Categories

- **Baseline Documents**
- **ISEC Studies**
- **Architecture**
- **Systems Engineering**
- **Management**
- **Tether Materials and Design**
- **Environmental [debris, radiation, O_x, electromagnetic]**
- **Tether Dynamics and Electrodynamics**
- **In the Atmosphere [Earth Port, Multi-Stage, HQ/POC]**
- **Tether Climbers Design and Power**
- **NODES [GEO, Apex Anchor, Gates & Centers]**
- **Lunar and mars Elevators**
- **Miscellaneous**

Baseline Documents

- Edwards, Bradley and Eric Westling, Space Elevator – A Revolutionary Earth-to-Space Transportation System, BC Edwards publishing, 2002.
- Edwards, B. and Laine, M. (2003), "The Space Elevator". Available at: http://www.mill-creek-systems.com/HighLift/chapter3.html
- Ishikawa, Yoji, The Space Elevator Construction Concept, Obayashi Corporation, 2013, IAC-13-D4.3.6.
- Raitt, David, Space Elevators: A History, ISEC Report 2017.
- Swan, P., Raitt, Swan, Penny, Knapman. International Academy of Astronautics Study Report, Space Elevators: An Assessment of the Technological Feasibility and the Way Forward, Virginia Edition Publishing Company, Science Deck (2013) ISBN-13: 978-2917761311
- Swan, P., David Raitt, Space Elevator – 15 Year Update, Journal of British Interplanetary Society, Vol 69, No 06/07, Dec 2016.
- Swan, P., David Raitt, John Knapman, Akira Tsuchida, Michael Fitzgerald, Yoji Ishikawa, Road to the Space Elevator Era, **Virginia Edition Publishing Company**, Science Deck (2019) ISBN-19: 978-0-9913370-3-3

- Swan, Peter, Robert "Skip" Penny, and Cathy Swan, Space Elevator Survivability – Space Debris Mitigation, Lulu.com, 2011.
- Penny, Robert. Swan, Peter, & Cathy Swan, "Space Elevator Concept of Operations," ISEC Position Paper #2012-1, International Space Elevator Consortium, Fall, 2013.
- Penny, R., P. Swan, C. Swan, J. Knapman, P. Glaskowsky, Design Considerations for Space Elevator Tether Climbers, ISEC Study Report, www.lulu.com, 2014
- Fitzgerald, M, R. Penny, P. Swan, C. Swan, Space Elevator Architectures and Roadmaps, ISEC Study Report, lulu.com, 2015
- Fitzgerald, Michael, Vern Hall, Cathy Swan, Peter Swan, Design Considerations for Space Elevator Apex Anchor and GEO Node, ISEC Study Report, lulu.com, 2017.
- Hall, Vern, R. Penny, P. Glaskowsky, S. Schaeffer, Design Considerations for Space Elevator Earth Port, ISEC Study Report, www.lulu.com, 2016.
- Knapman, Joh, P. Glaskowsky, D. Gleeson, V. Hall, D. Wright, M. Fitzgerald, P. Swan, Design Considerations for the Multi-Stage Space Elevator, ISEC Study Report, lulu.com, 2018.
- Swan, Peter, Michael Fitzgerald, "Space Elevator GEO Node and Apex Anchor Architectures," IAC-17, paper and presentation, Adelaide, Australia, Sept 2017.
- Semon, Ted Editor, "CLIMB Journal," Vol 1, No. 1, Lulu.com, Dec 2011.
- Semon, Ted Editor, "CLIMB Journal," Vol 2, No. 1, Lulu.com, July 2013.
- Semon, Ted Editor, "Via Ad Astra - The Space Elevator Magazine," Vol. 1, No. 1, Lulu.com, Dec 2015.
- Wright, Dennis, S. Avery, J. Knapman, M. Lades, P. Roubekas, P. Swan; Design Considerations for a Software Space Elevator Simulator, ISEC Study Report, lulu.com, 2017

Architecture

- Aleksandrov, Oleg, VERSION OF THE SPACE ELEVATOR, IAC-14, paper and presentation, Toronto, Oct 2014.
- Artsutanov, Y. (1960), "Into the Cosmos by Electric Rocket," Komsomolskaya Pravda, 31 July 1960. (Contents described in English, Lvov in Science, 158, 946-947, 1967.)
- Artsutanov, Y., "Into the Cosmos by Electric Rocket," Komsomolskaya Pravda, 31 July 1960. (The contents are translated by Lvov in Science, 158, 946-947, 1967.)
- Artsutanov, Y., "Into the Cosmos without Rockets," Znanije-Sila 7, 25, 1969.
- Artsutanov, Y., "Railway 'Moon-Earth'," Technika Molodishi, No. 4, p. 21, 1979.
- Cavallini, Anders, MEO Tethered Space Elevator System Architecture Feasibility Study, IAC-17, paper and presentation, Adelaide, Australia, Sept 2017.
- Edwards, Bradley and Eric Westling, Space Elevator – A Revolutionary Earth-to-Space Transportation System, BC Edwards publishing, 2002.
- Edwards, Bradley, The Space Elevator, NIAC Phase I Study Report, 2000.
- Edwards, B. The Space Elevator, NIAC Phase II Final Report, 2003.
- Fitzgerald, M, R. Penny, P. Swan, C. Swan, Space Elevator Architectures and Roadmaps, ISEC Study Report, lulu.com, 2015
- Fitzgerald, Michael, "Galactic Harbour, a Strategic Vision Emerges," Presentation at the National Space Society Conference, St. Louis, May 2017.
- Fitzgerald, Michael, "Space Elevator Sequences and Initial Operational Capability," Paper given at 2016 ISEC Space Elevator Conference, Seattle, 19-21 August 2016.
- Fitzgerald, Michael, "Strategic Approach to the Space Elevator," presented at 2017 ISEC Conference, Seattle, 25-27 Aug 2017.
- Fitzgerald, Michael, "Review of Architectural Notes," presented at 2017 ISEC Conference, Seattle, 25-27 Aug 2017.
- Fitzgerald, Michael, "Space Elevator Delineation," presented at 2017 ISEC Conference, Seattle, 25-27 Aug 2017.
- Fitzgerald, Michael, Preliminary Technology Readiness Assessment of the Space Elevator Transportation System. International Space Elevator Conference, paper and presentation, Seattle, Aug 2018.
- Fitzgerald, Michael, Final Summary and Brief on Preliminary Technology Readiness Assessment of the Space Elevator Transportation, System, International Space Elevator Conference, paper and presentation, Seattle, Aug 2018.
- Fitzgerald, Michael, Space Elevator Pathway to Technology Maturity … and Beyond, From Fountains to Tech Ready. presented at 2019 International Space Elevator Conference, Seattle, 16-18 Aug 2019.
- Fitzgerald, Michael, Pathway to Technological Maturity and Beyond. Presented at NSS International Space Development Conference, Washington, D.C. June 7-9 June, 2019.
- Fitzgerald, Michael, Technical Maturity and Development Readiness of the Galactic Harbour, IAC-19, paper and presentation, Washington D.C., Oct 2019

- Fitzgerald, Michael, Pathway Chronicles – Some Anecdotes, Architecture Notes as a Diary. presented at 2019 International Space Elevator Conference, Seattle, 16-18 Aug 2019.
- Gardner, J. (2003), "Where on Earth? Choosing an Anchor Point," 2nd Annual International Space Elevator Conference, Sante Fe, NM. Oct 2003
- Gassend, B. (2004), "Non-Equatorial Uniform-Stress Space Elevator," 3rd Annual International Space Elevator Conference, Washington DC, 20 June 2004.
- Gassend, B. (2004), "Exponential tethers for accelerated space elevator deployment". In Proc. of 3rd International Space Elevator Conference, June 2004.
- Isaacs, John, Allyn Vine, Hugh Bradner, George Bachus, "Satellite Elongation into a True "Sky-Hook," Science, Vol 151, Issue 3711, pg 682-683, 11 Feb 1966
- Ishikawa, Yoji, The Space Elevator Construction Concept, Obayashi Corporation, 2013, IAC-13-D4.3.6.
- Ishikawa Yoji, Obayashi Corporation's Space Elevator Construction Concept, Journal of British Interplanetary Society, Vol 69, No 06/07, Dec 2016.
- JSTM (2010), "Strategic Technology Road Map 2010", Ministry of Economy, Trade and Industry of Japan. Available
- Knapman, John, P. Glaskowsky, D. Gleeson, V. Hall, D. Wright, M. Fitzgerald, P. Swan, Design Considerations for the Multi-Stage Space Elevator, ISEC Study Report, lulu.com, 2018.
- at: http://www.meti.go.jp/policy/economy/gijutsu_kakushin/kenkyu_kaihatu/str2010.html (In Japanese only)
- Laine, M. (2006), "LiftPort Group Space Elevator Road Map." LiftPort, 2006
- Lang, D. D., "Space elevator initial construction mission overview", URL: http://home.comcast.net/~GTOSS/S (cited 1 Feb. 2010)
- Laubscher, Bryan, Space Elevator Systems Overview, International Space Elevator Conference, paper and presentation, Seattle, Aug 2018.
- Merrow, E. (2011), "Industrial Megaprojects, Concepts, Strategies, and Practices for Success". John Wiley & Sons, 2011
- METI. (2010). "Strategic Technology Roadmap". Ministry of Economy, Trade & Industry of Japan, 2010. Available (in Japanese only) at: http://www.meti.go.jp/policy/economy/gijutsu_kakushin/kenkyu_kaihatu/str2010.html
- METI. (2010) "Technology Strategy Map", Ministry of Economy, Trade and Industry of Japan, 2010. Available at: http://www.meti.go.jp/policy/economy/gijutsu_kakushin/kenkyu_kaihatu/str2010.html
- Nogawa, Yuichiro, Space Elevator Concept Comparison Summary, IAC-14, paper and presentation, Toronto, Oct 2014.
- Pasko, Vadym, Space Elevator. Alternative Design Solutions., IAC-15, paper and presentation, Jerusalem, Oct 2015.
- Pearson, Jerome, "The Orbital Tower: a Spacecraft Launcher using the Earth's rotational energy," Acta Astronautica, Vol 2, pp 785-799, Jan. 1975.
- Pearson, Jerome, "Using the Orbital Tower to Launch Earth-Escape Spacecraft Daily," IAF-76, October 1976.
- Pearson, J., E. Levin, J. Oldson, and H. Wykes, Lunar Space Elevators for CISLUNAR Space Development, NIAC Phase I Final Technical Report, 2 May 2005.
- Pullum, Laura, Space Elevator's Architectural View – 1, IAC-04, paper and presentation, Vancouver, Oct 2004

- Ragan, P. and B. Edwards, Leaving the Planet by Space Elevator, www.lulu.com, 2006.
- Shelef, B., "The Space Elevator Feasibility Condition", Climb Journal, Volume 1, Number 1, p. 87. And in - Spaceward Foundation, 2008. Available at: http://www.spaceward.org/elevator-library#SW
- Shelef, B., "Segment Based Ribbon Architecture"., In Proc. of 3rd International Space Elevator Conference, June 2004.
- Shelef, B., "A Solar-Based Space Elevator Architecture," Spaceward Foundation, 2008. http://www.spaceward.org/elevator-library#SW
- Squibb, Gael, Daryl Boden, and Wiley Larson, Cost Effective Space Mission Operations, McGraw Hill, 1996.
- Swan, P., Raitt, Swan, Penny, Knapman. International Academy of Astronautics Study Report, Space Elevators: An Assessment of the Technological Feasibility and the Way Forward, Virginia Edition Publishing Company, 2013.
- Swan, Peter, Safe Space Elevator – An Expectation to be Met Through a System Architecture Approach, IAC-04, paper and presentation, Vancouver, Oct 2004.
- Swan, Peter, Space Elevator Vision - An Enabler, IAC-06, paper and presentation, Valencia, Oct 2006
- Swan, Peter and Cathy Swan, Space Elevator Systems Architecture, Lulu.com publishers, 2007.
- Swan, Peter, "Space Elevator 101, Status and Architectures," Presentation at the National Space Society Conference, St. Louis, May 2017.
- Swan, Peter, Role of a Space Elevator Systems Architect, IAC-07, paper and presentation, Naples, Oct 2007.
- Swan, P., Space Elevator Current and Future Thrusts, Journal of British Interplanetary Society, Vol 69, No 06/07, Dec 2016.
- Swan, Peter, Michael Fitzgerald, "How the Space Elevator Grew into a Galactic Harbour," IAC-17, paper and presentation, Adelaide, Australia, Sept 2017.
- Swan, Peter Michael Fitzgerald, Galactic Harbour Duality – Enterprise and Infrastructure, IAC-18, paper and presentation, Bremen, Oct 2018.
- Swan, Peter, Apex Anchor Fast Transit to Mars. presented at 2019 International Space Elevator Conference, Seattle, 16-18 Aug 2019.
- Swan, P., David Raitt, John Knapman, Akira Tsuchida, Michael Fitzgerald, Yoji Ishikawa, Road to the Space Elevator Era, Virginia Edition Publishing Company, Science Deck (2019) ISBN-19: 978-0-9913370-3-3
- Torla, James and Matthew Peet, Space Elevator Support for Interplanetary Flight. Presented at NSS International Space Development Conference, Washington, D.C. June 7-9 June, 2019.
- TSM (2010), "Technological Strategy Zmap 2010 – Energy", Ministry of Economy, Trade and Industry. Available at: http://www.meti.go.jp/policy/economy/gijutsu_kakushin/kenkyu_kaihatu/str2010 download.html
- Tsuchida, Akira, "A Space Elevator Roadmap 2010," 2010 IAC, Prague, Oct 2010.
- Tsuchida, Akira, Space Elevator Road Map 2011, IAC-11, paper and presentation, Cape Town, Oct 2011.
- Tsuchida, Akira, Space Elevator Roadmap 2012, IAC-12, paper and presentation, Naples, Oct 2011.
- Tsuchida, Akira, Japanese Space Train concept, 2009 IAC, paper and presentation, Daejeon, Oct 2009.

- Tsuchida, Akira, et al. (2011), "Space Elevator Road Map 2011", 62nd International Astronautical Congress, Cape Town, Republic of South Africa, 2011
- Tsuchida A. et al (2009), "New Space Transportation System-Space Train (Elevator) : World trends and Japanese Space Train Concept",
- Tsuchida A. et al, "New Space Transportation System-Space Train (Elevator) : World trends and Japanese Space Train Concept", Technical report of IEICE. SANE 109(101), 93-98, 2009-06-18.
- Technical report of IEICE. SANE 109(101), 93-98, 2009-06-18
- USAF (2012), "Energy Horizons", United States Air Force, Energy S&T Vision 2011-2026, AF/ST TR 11-01 31 January 2012, Pgs. 21-24.
- Welch, J. (2012), http://thinkexist.com/quotation/good_business_leaders_create_a_vision-articulate/151585.html, June 2012.
- Wright, Dennis, S. Avery, J. Knapman, M. Lades, P. Roubekas, P. Swan; Design Considerations for a Software Space Elevator Simulator, ISEC Study Report, lulu.com, 2017

- AIAA/INCOSE. (1997) "SE Primer - Systems Engineering, A briefing", August 1997.
- Angel, R. (2006), "Feasibility of Cooling the Earth with a Cloud of Small Spacecraft near L1," Proceedings of the National Academy of Sciences, v 103, n46, 2006 November 14, 2006. Pp. 17184–17189. Available at: http://www.ncbi.nlm.nih.gov/pmc/articles/PMC1859907
- Aravind, P.K., The physics of the space elevator, Am. J. Phys., 75 (2007), pp. 125–130
- Bumgardner, Marvin, Optimization of Low-Thrust Orbit Transfer During Initial Ribbon Deployment for the Space Elevator, IAC-04, paper and presentation, Vancouver, Oct 2004.
- Chapman, P. K. (2010), "Deploying Sunsats", Online Journal of Space Communications, Issue 16, Winter 2010: Solar Power Satellites. Available at: http://spacejournal.ohio.edu/issue16/chapman.html
- Chati, Y.S., SPACE ELEVATOR: PHYSICAL PROPERTIES AND TRANSPORTATION SCENARIOS, CLIMB, Vol. I, 2011.
- Cohen, S. and Misra, A. K., "Satellite Placement Using the Space Elevator," Climb, Vol. 2, No. 1, 2013.
- Dempsey, J. SPACE ELEVATOR DEPLOYMENT, CLIMB, Vol. II, 2013.
- Edwards, Bradley and Eric Westling, Space Elevator – A Revolutionary Earth-to-Space Transportation System, BC Edwards publishing, 2002.
- Fitzgerald, M, R. Penny, P. Swan, C. Swan, Space Elevator Architectures and Roadmaps, ISEC Study Report, lulu.com, 2015
- Fitzgerald, Michael, "Space Elevator Sequences and Initial Operational Capability," Paper given at 2016 ISEC Space Elevator Conference, Seattle, 19-21 August 2016.
- Fitzgerald, Michael, Preliminary Technology Readiness Assessment of the Space Elevator Transportation System. International Space Elevator Conference, paper and presentation, Seattle, Aug 2018.
- Fitzgerald, Michael, Final Summary and Brief on Preliminary Technology REadiness Assessment of the Space Elevator Transportation, System, International Space Elevator Conference, paper and presentation, Seattle, Aug 2018.
- Fitzgerald, Michael, Pathway to Technological Maturity and Beyond. Presented at NSS International Space Development Conference, Washington, D.C. June 7-9 June, 2019.
- Fitzgerald, Michael, Technical Maturity and Development Readiness of the Galactic Harbour, IAC-19, paper and presentation, Washington D.C., Oct 2019
- Hall, Vern, R. Penny, P. Glaskowsky, S. Schaeffer, Design Considerations for Space Elevator Earth Port, ISEC Study Report, www.lulu.com, 2016
- Hein, Andreas, Producing a Space Elevator Tether using a NEO: A Preliminary Assessment, IAC-12, paper and presentation, Naples, Oct 2011.
- Hinton, G., SEVEN DEADLY ASSUMPTIONS ABOUT SPACE ELEVATORS, CLIMB, Vol. I, 2011.
- INCOSE. (2006), "Systems Engineering Handbook", v3, June 2006
- Ishikawa, Yoji, The Space Elevator Construction Concept, Obayashi Corporation, 2013, IAC-13-D4.3.6.

- Iwase, Satoshi, Comfortableness in Space Elevator — Physiological Challenge, IAC presentation and paper, IAC-10 Session D4.
- Keshmiri, M. and Misra, A.K., "On the deployment of a subsatellite in a space elevator system", 63rd International Astronautical Congress, Naples, Italy, October 2012, Paper No. IAC-12.D.4.3.8.
- Keshmiri, M. and Misra, A.K., "On the deployment of a subsatellite in a space elevator system", 63rd International Astronautical Congress, Naples, Italy, October 2012, Paper No. IAC-12.D.4.3.8.
- Knapman, John, Space Elevator Technology and Research, Journal of British Interplanetary Society, Vol 69, No 06/07, Dec 2016.
- Knapman, John, Technical Description of the Multi-Stage Space Elevator, International Space Elevator Conference, paper and presentation, Seattle, Aug 2018.
- Korn, Stanley, The Space Conveyor Could Revolutionize Space Travel. presented at 2019 International Space Elevator Conference, Seattle, 16-18 Aug 2019.
- Lang, David, SPACE ELEVATOR INITIAL CONSTRUCTION MISSION OVERVIEW, CLIMB, Vol. II, 2013.
- Lansdorp, Bas, Design of High-Tension Elastically Deforming Space Tether Deployer, IAC-04, paper and presentation, Vancouver, Oct 2004.
- Larson, Wiley., Space Mission Analysis and Design, Space Technology Library, Microcosm Press, 1999.
- Larson, W. et al. (2009) "Applied Space Systems Engineering", McGraw Hill, 2009.
- Larson, W. et al (2009), "Applied Space Systems Engineering", McGraw Hill, Boston, 2009. Pg. 304.
- Larson, Wiley, Doug Kirkpatrick, Jerry Sellers, L. Dale Thomas, and Dinish Verma, **Applied Space Systems Engineering,** McGraw Hill, 2009.
- Long, Bryan, "Approach towards governments to support this major systems of systems development," Presentation at the National Space Society Conference, St. Louis, May 2017.
- Laubscher, Bryan, Space Elevator Systems Engineering Analysis (LA-UR-04-1035), IAC-04, paper and presentation, Vancouver, Oct 2004
- Meulenberg, Andrew, LEO-based space-elevator development using available materials and technologies, 2009 IAC, paper and presentation, Daejeon, Oct 2009.
- Meulenberg, Andrew, sling-on-a-ring: a realizable space elevator to leo?, IAC-08, paper and presentation, Glasgow, Oct 2008.
- Penny, Robert. Swan, Peter, & Cathy Swan, "Space Elevator Concept of Operations," ISEC Position Paper #2012-1, International Space Elevator Consortium, Fall, 2013.
- Penny, R. and Jones, R. (1983), "A Model for Evaluation of Satellite Population Management Alternatives", AFIT Master's Thesis, 1983.
- Penny, R., P. Swan, C. Swan, J. Knapman, P. Glaskowsky, Design Considerations for Space Elevator Tether Climbers, ISEC Study Report, www.lulu.com, 2014
- Penny, Robert, Design Considerations for Geo Node, Apex Anchor and Communications Architecture ISEC Study underway 2017.
- Pullum, Laura, Systems Engineering for the Space Elevator – Complexity, IAC-05, paper and presentation, Fukuoka, Oct 2005.
- Roberts, Sophia Lee, Methodologies for Mitigating Risk to the Lower Reaches of the Future Space Elevators. presented at 2019 International Space Elevator Conference, Seattle, 16-18 Aug 2019.

- Robinson, Peter, Space Elevator Simulation: Validation and Metrology, Via Ad Astra, Vol 1, 2015.
- Robinson, Peter, Proposals for Growing Space Elevator TRL by operation of Demonstrator System. IAC-18, paper and presentation, Bremen, Oct 2018.
- Shelef, B. (2004), "Segment Based Ribbon Architecture"., In Proc. of 3rd International Space Elevator Conference, June 2004.
- Shelef, B. (2011), "The Space Elevator Feasibility Condition", Climb Journal, Volume 1, Number 1, p. 87.
- Shelef, B. (2008a), "Space Elevator Power System Analysis and Optimization, Spaceward Foundation, 2008. Available at: http://www.spaceward.org/elevator-library#SW
- Shelef, Ben, SPACE ELEVATOR POWER SYSTEM ANALYSIS AND OPTIMIZATION, CLIMB, Vol. II, 2013.
- Shelef, B. (2008b), "The Space Elevator Feasibility Condition", Spaceward Foundation, 2008. Available at: http://www.spaceward.org/elevator-library#SW
- Shelef, B., "The Space Elevator Feasibility Condition", Climb Journal, Volume 1, Number 1, p. 87. And in - Spaceward Foundation, 2008. Available at: http://www.spaceward.org/elevator-library#SW
- Shelef, Ben, ASTEROID SLINGSHOT EXPRESS - TETHER-BASED SAMPLE RETURN, CLIMB, Vol. II, 2013.
- Smitherman, David, Technology Development and Demonstration Concepts for the Space Elevator, IAC-04, paper and presentation, Vancouver, Oct 2004
- Sun, Sean and Dan Popescu, Building the Space Elevator: Lessons from Biological Design, International Space Elevator Conference, paper and presentation, Seattle, Aug 2018.
- Suzuki, Yuto, Suggestions of Research Areas and Future Experiments - Status Report of IAA SG3.24, IAC-17, paper and presentation, Adelaide, Australia, Sept 2017.
- Swan, Peter, Space Elevator Base Leg Architecture, IAC-04, paper and presentation, Vancouver, Oct 2004.
- Swan, Peter, Robert "Skip" Penny, and Cathy Swan, Space Elevator Survivability – Space Debris Mitigation, Lulu.com, 2011.
- Swan, P., Raitt, Swan, Penny, Knapman. International Academy of Astronautics Study Report, Space Elevators: An Assessment of the Technological Feasibility and the Way Forward, Virginia Edition Publishing Company, 2013.
- Swan, P., David Raitt, John Knapman, Akira Tsuchida, Michael Fitzgerald, Yoji Ishikawa, Road to the Space Elevator Era, Virginia Edition Publishing Company, Science Deck (2019) ISBN-19: 978-0-9913370-3-3
- Swan, P., Fitzgerald, M. "Space Elevator Development Sequence," IAC-16, paper and presentation, International Astronautical Congress, Guadalajara, Sept 2016,. IAC-16-D4.3.8.
- Swan, Peter, Characteristics of Space Elevator Apex Anchor, International Space Elevator Conference, paper and presentation, Seattle, Aug 2018.
- Swan, Peter, Interplanetary Mission Support from Galactic Harbour Apex Anchor, IAC-19, paper and presentation, Washington D.C., Oct 2019
- Swan, Peter, Apex Anchor Fast Transit to Mars. presented at 2019 International Space Elevator Conference, Seattle, 16-18 Aug 2019.
- Takahashi, Sakurako, Critical technologies for Space Elevator - Status report of IAA SG3.24, IAC-16, paper and presentation, Guadalajara, Oct 2016

- Tsuchida, Akira, Preliminary Systems Requirements for the Space Toilet on the Space Train, IAC presentation and paper, IAC-10 Session D4.
- Tsuchida, Akira, Non-technological risk abstraction and consideration for space elevator development, IAC-18, paper and presentation, Bremen, Oct 2018.
- Wertz, James R. (2011), "Space Mission Engineering: The New SMAD," Microcosm Press, Hawthorne Calif., 2011.
- Woo, Pamela, Energy Considerations in the Partial Space Elevator, IAC-13, paper and presentation, Beijing, Oct 2013.
- Wood, James, SEaCCIMT Proposal with Hollow Space Cable and Torque from Spinning Counterweight Wheels as Necessary Option to Create Multi-Stage Elevator, International Space Elevator Conference, paper and presentation, Seattle, Aug 2018.
- Wright, Dennis, A Hardware Space Elevator Simulator, International Space Elevator Conference, paper and presentation, Seattle, Aug 2018.
- Yamagiwa, Yoshiki, Verification of Space Elevator Technologies; Present Status and Future Plan in Japan, IAC-17, paper and presentation, Adelaide, Australia, Sept 2017.

Management

- Artsutanov, Y. (1960), "Into the Cosmos by Electric Rocket," Komsomolskaya Pravda, 31 July 1960. (Contents described in English, Lvov in Science, 158, 946-947, 1967.)
- Artsutanov, Y., "Into the Cosmos by Electric Rocket," Komsomolskaya Pravda, 31 July 1960. (The contents are translated by Lvov in Science, 158, 946-947, 1967.)
- Artsutanov, Y., "Into the Cosmos without Rockets," Znanije-Sila 7, 25, 1969.
- Artsutanov, Y., "Railway 'Moon-Earth'," Technika Molodishi, No. 4, p. 21, 1979.
- Angel, R. (2006), "Feasibility of Cooling the Earth with a Cloud of Small Spacecraft near L1," Proceedings of the National Academy of Sciences, v 103, n46, 2006 November 14, 2006. Pp. 17184–17189. Available at: http://www.ncbi.nlm.nih.gov/pmc/articles/PMC1859907
- Bernard-Cooper, Joshua, An Instrumented, Balloon-supported Tether for Early Space Elevator Research and Revenue. presented at 2019 International Space Elevator Conference, Seattle, 16-18 Aug 2019.
- Boada, Ivan, Our Answer to NASA's Beam Power Challenge, IAC-06, paper and presentation, Valencia, Oct 2006.
- Chapman, P. K. (2010), "Deploying Sunsats", Online Journal of Space Communications, Issue 16, Winter 2010: Solar Power Satellites. Available at: http://spacejournal.ohio.edu/issue16/chapman.html
- Chase, Frank, ISEC THEME POSTERS, CLIMB, Vol. II, 2013.
- Clarke, A. C. (1979), "The Fountains of Paradise", Harcourt Brace Jovanovich, New York, 1979.
- Cohen, S. and Misra, A. K., "Satellite Placement Using the Space Elevator," Climb, Vol. 2, No. 1, 2013.
- Cohen, Stephen and Arun Misra, SATELLITE PLACEMENT USING THE SPACE ELEVATOR, CLIMB, Vol. II, 2013.
- Dillon, S. (1892), "Historic moments: Driving the last spike of the Union Pacific". Scribner's Magazine, August 1892. pp. 25-259. Available at: http://www.unz.org/Pub/Scribners-1892aug-00253
- Dodrill, Mark, The ISEC History – Interview with Vern Hall, Via Ad Astra, Vol 1, 2015.
- Edwards, B, et.al., The Space Elevator, NIAC Study – the NASA Institute for Advanced Concepts Phase I, Oct. 2000. http://www.niac.usra.edu/studies/472Edwards.html
- Edwards, B., et.al., The Space Elevator, NIAC Phase II Jan. 2003. http://www.niac.usra.edu/studies/521Edwards.html.)
- Edwards, Bradley, The Space Elevator and NASA's New Space Initiative, IAC-04, paper and presentation, Vancouver, Oct 2004.
- Edwards, Bradley, The Space Elevator Program at ISR, IAC-04, paper and presentation, Vancouver, Oct 2004.
- Edwards, Bradley, Private Investment and Space Elevator Development Activities, IAC-05, paper and presentation, Fukuoka, Oct 2005.
- Edwards, B. and Ragan, P. (2006), Leaving the Planet by Space Elevator, Lulu.com

- Edwards, Bradley and Eric Westling, Space Elevator – A Revolutionary Earth-to-Space Transportation System, BC Edwards publishing, 2002.
- Elkins-Tanton, Linda, KEYNOTE: The ASU Interplanetary Initiative: Advancing Society Through Exploration. IAC-19, paper and presentation, Washington D.C., Oct 2019
- Fitzgerald, Michael, Space Elevator Pathway to Technology Maturity … and Beyond, From Fountains to Tech Ready. presented at 2019 International Space Elevator Conference, Seattle, 16-18 Aug 2019.
- Forest, Kevin, "New ISEC Website – Space Elevator Home," presented at 2017 ISEC Conference, Seattle, 25-27 Aug 2017.
- Gilbertson, R., FINDING AND TRANSLATING ARTSUTANOV'S ORIGINAL SPACE ELEVATOR ARTICLE FROM 1960, CLIMB, Vol. I, 2011.
- Harris, R. (2012), "Rio Environment Meeting Focuses On 'Energy for All'" 19 June 2012. Available at: http://www.wbur.org/npr/155294726/rio-environment-meeting-focuses-on-energy-for-all
- Heckman, Fuller-Clarke Sphere, International Space Elevator Conference, paper and presentation, Seattle, Aug 2018.
- Isaacs, J., Vine, A. C., Bradner, H. and Bachus, G. E. (1966), "Satellite Elongation into a true Sky-Hook," Science, 151, 682-683, 1966.
- Ishikawa, Yoji, The Space Elevator Construction Concept, Obayashi Corporation, 2013, IAC-13-D4.3.6.
- Kai, Sunao, The Law of the Space elevator -- The relationship to the Law of the Space, the Sea and the Sky, IAC-12, paper and presentation, Naples, Oct 2011.
- Kai, Sunao, "Law and Structure of the Space Elevator," presented at 2017 ISEC Conference, Seattle, 25-27 Aug 2017.
- Kirchner, Stefan, International Law and the Construction and Operation of a Tethered Space Elevator, Via Ad Astra, Vol 1, 2015.
- Knapman, John, Space Elevator Research, IAC-14, paper and presentation, Toronto, Oct 2014.
- Laubscher, Bryan, The Space Elevator and Planetary Defense, IAC-05, paper and presentation, Fukuoka, Oct 2005.
- Laubscher, Bryan, "Space Elevator System Overview," presented at 2017 ISEC Conference, Seattle, 25-27 Aug 2017.
- Laubscher, Bryan, Space Elevator Systems Overview, International Space Elevator Conference, paper and presentation, Seattle, Aug 2018.
- Loubeyre, R., QUESTIONING THE SPACE ELEVATOR LEGAL RISK MANAGEMENT REGIME, CLIMB, Vol. I, 2011.
- Mankins, J. (2011), "Space Solar Power, The First International Assessment Of Space Solar Power: Opportunities, Issues And Potential Pathways Forward", IAA, October 2011.
- Martin, Nichlos, Early Space Elevator History – Tsiolkovsky, Artsutanov and Pearson, Via Ad Astra, Vol 1, 2015.
- Matloff, Gregory, The Partial Space Beanstalk: Its Application to Space Migration and Commerce, IAC-08, paper and presentation, Glasgow, Oct 2008.
- Nelson, Paul, Releasing Earth Space Elevator Climbers into Geostationary Orbit, IAC-05, paper and presentation, Fukuoka, Oct 2005.
- NTRM (2012), "NASA Space Technology Roadmaps and Priorities: Restoring NASA's Technological Edge and Paving the Way for a New Era in Space," National Academy of Science Report, Washington D.C., 2012.

- National Space Society. (2007), "Space Solar Power: An Investment for Today – An Energy Solution for Tomorrow," National Space Society, Oct. 2007
- Pearson, J., "The Orbital Tower: A Spacecraft Launcher Using the Earth's Rotational Energy," Acta Astronautica, Vol. 2, pp. 785-799, Sep/Oct 1975.
- Pearson, Jerome, The real history of the space elevator, IAC-06, paper and presentation, Valencia, Oct 2006.
- Pearson, Jerome, THE REAL HISTORY OF THE SPACE ELEVATOR, CLIMB, Vol. II, 2013.
- Pearson, Jerome, Space Elevators, Apollo, and Back to the Moon. Presented at NSS International Space Development Conference, Washington, D.C. June 7-9 June, 2019.
- Penny, Robert. Swan, Peter, & Cathy Swan, "Space Elevator Concept of Operations," ISEC Position Paper #2012-1, International Space Elevator Consortium, Fall, 2013.
- Penny, Robert, Space Elevator CONOPS Initial Thinking, IAC-12, paper and presentation, Naples, Oct 2011.
- Purang, Deepak (n.d.), "Space sunshade may one day reduce global warming." Editorial. Available at: http://www.streetdirectory.com/travel_guide/14921/gadgets/space_sunshade_may_one_day_reduce_global_warming.html
- Raitt, David, Space Elevators: A History, ISEC Study Report, May 2017.
- Raitt, David, The Space Elevator: Economics and Applications, IAC-04, paper and presentation, Vancouver, Oct 2004.
- Raitt, David, The Space Elevator: Historical and Future Perspectives, IAC-05, paper and presentation, Fukuoka, Oct 2005.
- Raitt, D. (2005), "The Space Elevator: its Place in History, Literature and the Arts. In: Proceedings of 56th International Astronautics Congress, 17-21 October 2005, Fukuoka, Japan. IAC, 2005. IAC-05-D4.3.02
- Raitt, D. and Edwards, B. (2004), "The Space Elevator: Economics and Applications." In: Proceedings of 55th IAC, 4-8 October 2004, Vancouver, Canada. IAC, 2004. IAC-04-IAA.3.8.3
- Raitt, David - "A Very Short History of Space Elevators," presented at 2017 ISEC Conference, Seattle, 25-27 Aug 2017.
- Sato, Minoru, Utilization of Space Elevator in Education and Outreach, IAC-17, paper and presentation, Adelaide, Australia, Sept 2017.
- Schlusser, Eugene and Natalie Sherman, Conversations with Yuri Artsutanov, Via Ad Astra, Vol 1, 2015.
- Semon, Ted, A brief history of the Space Elevator Games, Via Ad Astra, Vol 1, 2015.
- Smith, C. M. (2013), "Starship Humanity," Scientific American, Jan 2013, pg. 39-43.
- Smitherman, David, Critical Technologies for the Development of Future Space Elevator Systems, IAC-05, paper and presentation, Fukuoka, Oct 2005.
- Smitherman, D. (2006), "Space Elevators: An Advanced Earth-Space Infrastructure for the New Millennium", University Press of the Pacific, 2006.
- Smitherman, David, Earth-Based Space Elevator Research and Technology Development, IAC-06, paper and presentation, Valencia, Oct 2006.
- Squibb, Gael, Daryl Boden, and Wiley Larson, Cost Effective Space Mission Operations, McGraw Hill, 1996.

- Swan, Peter, WHY - The Motivation for a Space Elevator, IAC-05, paper and presentation, Fukuoka, Oct 2005.
- Swan, Peter, Yearly Study Reports from the International Space Elevator Consortium, Via Ad Astra, Vol 1, 2015.
- Swan, Peter, Space Elevator Lexicon, IAC-16, paper and presentation, Guadalajara, Oct 2016
- Swan, Peter, Cosmic Study Overview – Space Elevator Feasibility, IAC-12, paper and presentation, Naples, Oct 2011.
- Swan, Peter, Quick-Look Operations Concept for a Space Elevator, IAC-11, paper and presentation, Cape Town, Oct 2011.
- Swan, Peter, Creation of the International Space Elevator Consortium, 2009 IAC, paper and presentation, Daejeon, Oct 2009.
- Swan, Peter, Space Elevator Development Sequence, IAC-16, paper and presentation, Guadalajara, Oct 2016
- Swan, P., Raitt, Swan, Penny, Knapman. International Academy of Astronautics Study Report, Space Elevators: An Assessment of the Technological Feasibility and the Way Forward, Virginia Edition Publishing Company, 2013.
- Swan, Peter, "ISEC President's Corner," presented at 2017 ISEC Conference, Seattle, 25-27 Aug 2017.
- Swan, Peter, "ISEC Activision - The strength of our Volunteers," presented at 2017 ISEC Conference, Seattle, 25-27 Aug 2017.
- Swan, Peter, Interplanetary Mission Support from Galactic Harbour Apex Anchor, IAC-19, paper and presentation, Washington D.C., Oct 2019
- Swan, Peter Space Elevators are Closer Than You Think. Presented at NSS International Space Development Conference, Washington, D.C. June 7-9 June, 2019.
- Swan, Peter, The Future is Exciting for Space Elevators Presented at NSS International Space Development Conference, Washington, D.C. June 7-9 June, 2019.
- Swan, Phil, One Step Closer to Reality - Achieving Technical and Economic Viability through Innovative Business Models, International Space Elevator Conference, paper and presentation, Seattle, Aug 2018.
- Swan, Phil, Myths Busted: The Real Reason Launch Costs are High and How Space Access Infrastructure Can Reduce Launch Costs to LEO. presented at 2019 International Space Elevator Conference, Seattle, 16-18 Aug 2019.
- Torla, James, Matthew Peet, Optimization of Low Fuel and Time-Critical Interplanetary Transfers using Space Elevator Apex Anchor Release: Mars, Jupiter and Saturn. IAC-19, paper and presentation, Washington D.C., Oct 2019
- Tsiolkovsky, K. E. (1959), "Speculations of Earth and Sky and On Vesta", Moscow, USSR Academy of Sciences, 1959 (in Russian – first published 1895).
- Verge, Munir, International Cooperation and the Space Elevator, IAC-04, paper and presentation, Vancouver, Oct 2004.
- Wiese, Tim, A Journey of Student Space Elevator Development. IAC-19, paper and presentation, Washington D.C., Oct 2019
- Wittkotter, Erland, The Gamification of Space Elevator Technology. presented at 2019 International Space Elevator Conference, Seattle, 16-18 Aug 2019.
- Wright, Dennis, Space Elevator Baseline System Overview, presented at 2019 International Space Elevator Conference, Seattle, 16-18 Aug 2019.

Tether Materials and Design

Material Properties

- Artyukhov, V. / Liu, Y. / Yakobson, B., GETTING THE MOST OUT OF NANOTUBES: GUIDANCE FROM FRACTURE PHYSICS AND ATOMISTIC SIMULATIONS, CLIMB, Vol. II, 2013.
- Barber, A.H. et al. (2005), "Stochastic strength of nanotubes: An appraisal of available data", Compos. Sci. Technol., Vol 65 No 15-16, pp. 2380–2384.
- Barnds, J., et al. (1998). TiPS: Results of a Tethered Satellite System. Tether technology Interchange Meeting, NASA/CP-1998-206900, NASA Marshall. January 1998.
- Belytschko, T. et al (2002), "Atomistic simulations of nanotube fracture", Phys. Rev. B, Vol 65 no 23, 235430.
- Brambilla, G. & Payne, D.N. (2009) "The ultimate strength of glass silica nanowires", Nano Lett., Vol 9 No 2, pp. 831-835.
- Brambilla, G., AN UPDATED REVIEW OF NANOTECHNOLOGIES FOR THE SPACE ELEVATOR TETHER, CLIMB, Vol. I, 2011.
- Carroll, J.A. (1993). SEDS Deployer Design and Flight performance. AIAA Space Programs and Technologies Conference and Exhibit, Huntsville, AIAA-93-4764. September 1993.
- Carroll, J.A. and Oldson, J.C. (1995). SEDS characteristics and capabilities. In Proceedings of the 4th International Conference on Tethers in Space, pp. 1079-1090.
- Chobotov, V.A. and Mains, D.L. (1999). Tether Satellite System Collision Study, , Acta Astronautica, Vol 44, Nos. 7 – 12, pp 543 – 551, 1999.
- Cornwell, C.F. et al. (2011) "Very-high-strength (60-GPa) carbon nanotube fiber design based on molecular dynamics simulations", J. Chem. Phys., Vol. 134, 204708.
- Cosmo, M.L., and Lorenzini, C.E., Tethers In Space Handbook, prepared for NASA Marshall Space Flight Center by Smithsonian Astrophysics Observatory, Cambridge, MA, December 1977.
- Coste, Darren, Tether Material and Deployment Mechanism for Tethered Satellite Cons\tellations\IAC-18, paper and presentation, Bremen, Oct 2018.
- Cronin, S.B. et al (2005), "Resonant Raman spectroscopy of individual metallic and semiconducting single-wall carbon nanotubes under uniaxial strain", Phys. Rev. B, Vol 72 No 3, 035425.
- Demczyk, B.G. et al (2002), "Direct mechanical measurement of the tensile strength and elastic modulus of multiwalled carbon nanotubes", Materials Science and Engineering, vol A334, pp. 173–178.
- Dobrowolny, M. and Stone, N.H. (1994). A Technical Overview of TSS-1: The First Tethered Satellite System Mission, Il Nuovo Cimento, Vol. 17C, N.1, pp. 1-12.
- Ebbesen, T.W. & Ajayan, P.M. (1992) "Large-scale synthesis of carbon nanotubes" Nature, Vol. 358, pp. 220-222.
- Filleter, T. & Espinosa, H.D. (2012) "Multi-scale mechanical improvement produced in carbon nanotube fibers by irradiation cross-linking", Carbon, in press (2013)

- Gao, P. et al (2010), "Self-Built Tensile Strain in Large Single-Walled Carbon Nanotubes", ACS Nano, Vol 4 No 2, pp. 992-998.
- Gassend, B. (2004), "Non-Equatorial Uniform-Stress Space Elevator," 3rd Annual International Space Elevator Conference, Washington DC, 20 June 2004.
- Gassend, B. (2004), "Exponential tethers for accelerated space elevator deployment". In Proc. of 3rd International Space Elevator Conference, June 2004.
- Haase, Mark, Advances in High Tensile Strength Materials for Space Elevator Applications, Journal of British Interplanetary Society, Vol 69, No 06/07, Dec 2016.
- Haase, Mark, "Status of Space Elevator Tether Material Research," Presentation at the National Space Society Conference, St. Louis, May 2017.
- Haase, Mark, "Status of Space Elevator Tether Material Research," Presented during the Space Elevator Track at the 2017 National Space Society's International Space Development Conference, St. Louis, August 2017.
- Hata, K. et al. (2004) "Water-Assisted Highly Efficient Synthesis of Impurity-Free Single-Walled Carbon Nanotubes" Science, Vol. 306 no 5700, pp. 1362-1364.
- Huang, S. et al (2004) "Growth Mechanism of Oriented Long Single Walled Carbon Nanotubes Using "Fast-Heating" Chemical Vapor Deposition Process", Nano Lett., Vol 4 No 6, pp. 1025-1028.
- Hong, B. et al. (2005) "Quasi-Continuous Growth of Ultralong Carbon Nanotube Arrays", J. Am. Chem. Soc., Vol. 127 No 44, pp. 15336-15337.
- Iijima, S. (1991) "Helical microtubules of graphitic carbon" Nature, Vol. 354, pp. 56-58.
- Ishikawa, Yoji, Survivabilitiy of Carbon Nanotubes in Space, IAC-18, paper and presentation, Bremen, Oct 2018.
- Iwanaga, H. & Kawai, C. (2005) "Tensile Strength of Silicon Nitride Whiskers Synthesized by Reacting Amorphous Silicon Nitride and Titanium Dioxide", J. Am. Ceram. Soc., Vol 81 No 3, pp. 773-776.
- Kong, J. et al. (1998) "Synthesis of individual single-walled carbon nanotubes on patterned silicon wafers" Nature, Vol. 395, pp. 878-881.
- Koziol, K. et al (2007) "High-Performance Carbon Nanotube Fiber", Science, Vol 318 No 5858, pp. 1892-1895
- Kruijff, M. and Heide, E. J. van der. (2009). Qualification and In-flight Demonstration of a European Tether Deployment System on YES2. In Acta Astronautica, vol. 64, p.882-905.
- Kruijff M., Heide E.J. van der, Ockels W.J. (2009). Data Analysis of a Tethered SpaceMail Experiment. In Journal of Spacecraft and Rockets, Vol. 46, No. 6, pp. 1272-1287. (presented as AIAA-2008-7385).
- Laubscher, Bryan, "Carbon Nanotube Industry Review," presented at 2017 ISEC Conference, Seattle, 25-27 Aug 2017.
- Li, W.Z. et al. (1996) "Large-Scale Synthesis of Aligned Carbon Nanotubes" Science, Vol. 274 no 5293, pp. 1701-1703.
- Li, Y-L et al (2004) "Direct Spinning of Carbon Nanotube Fibers from Chemical Vapor Deposition Synthesis", Science, Vol 304 no 5668, pp 276-278.
- Lin, W. et al (2010) "Microwave Makes Carbon Nanotubes Less Defective", ACS Nano, Vol 4 No 3, pp. 1716-1722.
- Liu, K. et al (2010) "Carbon nanotube yarns with high tensile strength made by a twisting and shrinking method", Nanotech., Vol 21 No 4, 045708.

- Ma, W.J. et al (2007) "Directly Synthesized Strong, Highly Conducting, Transparent Single-Walled Carbon Nanotube Films", Nano Lett., Vol 7 No 8, pp. 2307-2311.
- Ma, W.J. et al (2009) "Monitoring a Micromechanical Process in Macroscale Carbon Nanotube Films and Fibers", Adv. Mat., Vol 21 No 5, 603-608.
- Nixon, Adrian, Keynote Presentation: The last piece of the puzzle? - Single Crystal Graphene, International Space Elevator Conference, paper and presentation, Seattle, Aug 2018.
- Nixon, Adrian, Update on Graphene as a Tether Material. presented at 2019 International Space Elevator Conference, Seattle, 16-18 Aug 2019.
- O'Brien, N.P. et al. (2012) "A theoretical quantification of the possible improvement in the mechanical properties of carbon nanotube bundles by carbon ion irradiation", Carbon, Vol. 53, pp. 346-356.
- Pan, Z. W. et al. (1998) "Very long carbon nanotubes". Nature, Vol. 394 pp 631-632.
- Peng, B. et al. (2008) "Measurements of near-ultimate strength for multiwalled carbon nanotubes and irradiation-induced crosslinking improvements", Nat. Nanotechol., Vol. 3 No 10, pp. 626-631.
- Pugno, N.M. & Ruoff, R.S. (2004) "Quantized fracture mechanics", Philosophical Mag., Vol 84 No 27, pp. 2829-2845.
- Pugno, N. et al (2009) "Size effects on the strength of nanotube bundles", Meas. Sci. Technol., Vol 20 No 8, 084028
- Pugno, N. M. (2013), Towards the Artsutanov's dream of the space elevator: The ultimate design of a 35 GPa strong tether thanks to graphene. Acta Astronautica, Volume 82, Issue 2, p. 221-224.
- Pugno, N., THE EFFECT OF COLLAPSED NANOTUBES ON NANOTUBE BUNDLE STRENGTH, CLIMB, Vol. II, 2013.
- Pugno, N. / Abdalrahman T., MODELING THE SELF-HEALING OF BIOLOGICAL OR BIO-INSPIRED NANOMATERIALS, CLIMB, Vol. I, 2011.
- Rentein, Peter, A Scalable Carbon Nanotube Cable Strengthening Method. presented at 2019 International Space Elevator Conference, Seattle, 16-18 Aug 2019.
- Roundy, D., and Cohen, M.L. (2001), "Ideal strength of diamond, Si, and Ge", Phys. Rev. B, vol. 64, 212103.
- Ruoff, R.S, Qian, D. & Liu, W.K. (2003) "Mechanical properties of carbon nanotubes: theoretical predictions and experimental measurements", Comptes Rendus Physique, Vol 4 No 9, pp. 993-1008
- Sasaki, S. et al (1987). Results from a Series of Tethered Rocket Experiments. AIAA, USA.
- Sasaki, S. and Oyama, K.I. (1994). Space Tether Experiments in Japan. 2nd International Workshop on the Application of Tethered Systems in Space, Kanagawa, Japan, ISAS, May 1994.
- Sharma, Gaurav and Andrew Meulenberg "Collosal Carbon Tubes as Tethers for a Space Elevator," Conf. Proceedings, International Space Elevator Consortium, Seattle, Washington, USA, 23 Aug. 2012
- Stano, K.L. et al (2008) "Direct spinning of carbon nanotube fibres from liquid feedstock", Int. J. Mat. Forming, Vol 1 No 2, pp. 59-62
- Suemori, K. (2012), "Film-shaped thermoelectric conversion elements can be produced in print" Available at: http://www.aist.go.jp/aist_j/aistinfo/aist_today/vol12_04/p17.html

- Swan, P., David Raitt, John Knapman, Akira Tsuchida, Michael Fitzgerald, Yoji Ishikawa, Road to the Space Elevator Era, **Virginia Edition Publishing Company,** Science Deck (2019) ISBN-19: 978-0-9913370-3-3

- Tang, G. et al (2010) "New Confinement Method for the Formation of Highly Aligned and Densely Packed Single-Walled Carbon Nanotube Monolayers", Small, Vol 6 No 14, pp. 1488-1491.

- Telling, R. H., Pickard, C. J., Payne, M. C., and Field, J. E. (2000), "Theoretical Strength and Cleavage of Diamond", Phys. Rev. Lett., Vol. 84, pp. 5160–5163.

- Termonia, Y. et al (1985), "Theoretical Study of the Influence of the Molecular Weight on the Maximum Tensile Strength of Polymer Fibers", Macromol., Vol. 18, pp. 2246-2252.

- Tsukiyama, Y. (2010)." Tribological properties of high-alignment carbon nanotube films", The Machine Design and Tribology Division meeting in JSME 2010 (10), 49-50, 2010-04-18

- Tyc, G. and Han, R.P.S. (1995). Attitude Dynamics Investigation of the OEDIPUS: A Tethered Rocket Payload. Journal of Spacecraft and Rockets, Vol. 32, No. 1, p. 133-141, February 1995.

- Umehara, N. (2007), "Tribology of Fullerane and Carbon NanoTube as Advanced Materials Designed Nano-structures", Shinku, Vol. 50 No. 2, 2007. Pp. 76-81

- Vigneron, F.R., Jablonski, A.M. et al. (1997). Comparison of Analytical Modeling of OEDIPUS Tethers with Data from Tether Laboratory. Journal of Guidance, Control and Dynamics, Vol. 20, No. 3, pp.471-478, May-June 1997.

- Wang, X. et al. (2009) "Fabrication of Ultralong and Electrically Uniform Single-Walled Carbon Nanotubes on Clean Substrates" Nano Lett. Vol. 9 no 9, pp. 3137-3141.

- Wang, Z., Ciselli, P. & Peijs, T. (2007), "The extraordinary reinforcing efficiency of single-walled carbon nanotubes in oriented poly(vinyl alcohol) tapes", IOP Nanotechnology, Vol 18 No 45, 455709.

- Wei, X. et al (2010) "Tensile Tests on Individual Multi-Walled Boron Nitride Nanotubes", Adv. Mater., Vol 22 No 43, pp. 4895-4899.

- Wen, Q. et al (2010) "Growing 20 cm Long DWNTs/TWNTs at a Rapid Growth Rate of 80−90 μm/s", Chem. Mater., Vol 22 No 4, pp. 1294-1296.

- Wen, Q. et al (2010b) "100-mm Long, Semiconducting Triple-Walled Carbon Nanotubes", Adv. Mater., Vol 22 No 16, pp. 1867-1871.

- Williams, P., Blanksby, C., Trivailo, P., "Tethered Planetary Capture Maneuvers," Journal of Spacecraft and Rockets, Vol. 41, No. 4, pp.603-613, 2004.

- Williams, P., "Dynamics and Control of Spinning Tethers for Rendezvous in Elliptic Orbits," Journal of Vibration and Control, Vol. 12, No. 7, pp.737-771, 2006.

- Wong, S.S. et al. (1997) "Nanobeam Mechanics: Elasticity, Strength, and Toughness of Nanorods and Nanotubes", Science, Vol 277 No 5334, pp 1971-1975.

- Xie, H. et al., "Growth of high-density parallel arrays of ultralong carbon nanotubes with catalysts pinned by silica nanospheres", Carbon, Vol. 52, pp. 535-540.

- Yakobson, B.I., & Avouris, P. (2001), "Mechanical properties of carbon nanotubes" Carbon Nanotubes, Vol 80, pp. 287–327.

- Yamada, T. et al (2008) "Revealing the Secret of Water-Assisted Carbon Nanotube Synthesis by Microscopic Observation of the Interaction of Water on the Catalysts", Nano Lett., Vol 8 No 12, pp. 4288-4292.
- Yao, Y. et al (2007) "Raman Spectral Measuring of the Growth Rate of Individual Single-Walled Carbon Nanotubes", J. Phys. Chem. C, Vol 111 No 24, pp. 8407-8409.
- Yu, M.-F. et al (2000) "Strength and Breaking Mechanism of Multiwalled Carbon Nanotubes Under Tensile Load", Science, Vol 287 No 5453, pp. 637-640.
- Yuan, Q. et al (2011) "Threshold Barrier of Carbon Nanotube Growth", Phys. Rev. Lett., Vol. 107, 156101.
- Yuan, Q. et al (2012) "Efficient Defect Healing in Catalytic Carbon Nanotube Growth", Phys. Rev. Lett., Vol. 108, 245505.
- Zedd, M.F. (1998). Experiments in Tether Dynamics Planned for ATEx's Flight. Tether technology Interchange Meeting, NASA/CP-1998-206900, NASA Marshall, January 1998.
- Zhang, M. et al (2004) "Multifunctional Carbon Nanotube Yarns by Downsizing an Ancient Technology", Science, Vol 306 No 5700, pp 1358-1361.
- Zhang, X. et al (2007) "Ultrastrong, Stiff, and Lightweight Carbon-Nanotube Fibers", Adv. Mater., Vol 19, pp. 4198-4201.
- Zhang, X. & Li, Q. (2010) "Enhancement of Friction between Carbon Nanotubes: An Efficient Strategy to Strengthen Fibers", ACS Nano, Vol 4 No 1, pp. 312-316.
- Zhang, R. et al (2011) "Superstrong Ultralong Carbon Nanotubes for Mechanical Energy Storage", Adv. Mater., Vol 23 No 30, pp. 3387-3391.
- Zhao, Q. et al (2002) "Ultimate strength of carbon nanotubes: A theoretical study", Phys. Rev. B, Vol 65 No 14, 144105.
- Zheng, L.X. et al (2004) "Ultralong single-wall carbon nanotubes", Nature Mat., Vol 3 No 10, pp. 673-676.
- Zheng, L. et al (2009) "Tuning Array Morphology for High-Strength Carbon-Nanotube Fibers", Small Vol 6 No 1, pp. 132-137.

Tether Design

- Ambartsumian, S.A. / Belubekyan, M.V. / Ghazaryan K.B., OPTIMAL DESIGN OF THE SPACE ELEVATOR TETHER, CLIMB, Vol. I, 2011.
- Barnds, J., et al. (1998). TiPS: Results of a Tethered Satellite System. Tether technology Interchange Meeting, NASA/CP-1998-206900, NASA Marshall. January 1998.
- Carroll, J.A. (1993). SEDS Deployer Design and Flight performance. AIAA Space Programs and Technologies Conference and Exhibit, Huntsville, AIAA-93-4764. September 1993.
- Carroll, J.A. and Oldson, J.C. (1995). SEDS characteristics and capabilities. In Proceedings of the 4th International Conference on Tethers in Space, pp. 1079-1090.
- Chobotov, V.A. and Mains, D.L. (1999). Tether Satellite System Collision Study, Acta Astronautica, Vol 44, Nos. 7 – 12, pp 543 – 551, 1999.
- Cosmo, M.L., and Lorenzini, C.E., Tethers in Space Handbook, prepared for NASA Marshall Space Flight Center by Smithsonian Astrophysics Observatory, Cambridge, MA, December 1977.
- Dobrowolny, M. and Stone, N.H. (1994). A Technical Overview of TSS-1: The First Tethered Satellite System Mission, Il Nuovo Cimento, Vol. 17C, N.1, pp. 1-12.
- Kruijff, M. and Heide, E. J. van der. (2009). Qualification and In-flight Demonstration of a European Tether Deployment System on YES2. In Acta Astronautica, vol. 64, p.882-905.
- Kruijff M., Heide E.J. van der, Ockels W.J. (2009). Data Analysis of a Tethered SpaceMail Experiment. In Journal of Spacecraft and Rockets, Vol. 46, No. 6, pp. 1272-1287. (presented as AIAA-2008-7385).
- Okada, Morihiro, Light and strong CNT fiber spun with CNT web, IAC presentation and paper, IAC-10 Session D4.
- Sasaki, S. et al (1987). Results from a Series of Tethered Rocket Experiments. AIAA, USA.
- Sasaki, S. and Oyama, K.I. (1994). Space Tether Experiments in Japan. 2nd International Workshop on the Application of Tethered Systems in Space, Kanagawa, Japan, ISAS, May 1994.
- Tyc, G. and Han, R.P.S. (1995). Attitude Dynamics Investigation of the OEDIPUS: A Tethered Rocket Payload. Journal of Spacecraft and Rockets, Vol. 32, No. 1, p. 133-141, February 1995.
- Vigneron, F.R., Jablonski, A.M. et al. (1997). Comparison of Analytical Modeling of OEDIPUS Tethers with Data from Tether Laboratory. Journal of Guidance, Control and Dynamics, Vol. 20, No. 3, pp.471-478, May-June 1997.
- Williams, P., Blanksby, C., Trivailo, P., "Tetehred Planetary Capture Maneuvers," Journal of Spacecraft and Rockets, Vol. 41, No. 4, pp.603-613, 2004.
- Williams, P., "Dynamics and Control of Spinning Tethers for Rendezvous in Elliptic Orbits," Journal of Vibration and Control, Vol. 12, No. 7, pp.737-771, 2006.

- Barry, R.G., Chorley, R.J. (1998), Atmosphere, Weather & Climate, (Seventh Edition), Routledge, London, Section 6-3
- Edwards, Bradley and Eric Westling, Space Elevator – A Revolutionary Earth-to-Space Transportation System, BC Edwards publishing, 2002.
- Hussey, John ed., Paper on Space Debris Mitigation Guidelines for Spacecraft, Draft_– International Academy of Astronautics, 2003.
- International Academy of Astronautics. (2000), "2001 Position Paper on Orbital Debris", International Academy of Astronautics, 24.11.2000.
- International Academy of Astronautics (2005), "2006 Position Paper on Space Debris Mitigation", International Academy of Astronautics, 10.15.2005.
- Ishige, Yuuki & Satorni Kawamoto. "Study on Electrodynamic Tether System for Space Debris Removal." (IAF-02-A.7.04.) 53rd International Astronautical Congress, 2002.
- Ishikawa, Yoji, The Space Elevator Construction Concept, Obayashi Corporation, 2013, IAC-13-D4.3.6.
- Jorgensen, A., B. Gassend, R. H. W. Friedel, T. Cayton, S. E. Patamia, "Space Elevator Radiation Hazards and How to Mitigate Them," 3rd Space Elevator Conference, Washington, DC, June 29, 2004.
- Jorgensen, Anders, Active radiation shielding for the proposed space elevator using magnetic coils, IAC-06, paper and presentation, Valencia, Oct 2006.
- Jorgensen, A. M., S. E. Patamia, and B. Gassend. "Passive radiation shielding considerations for the proposed space elevator." Acta Astronautica 60.3 (2006): 198-209.
- Jorgensen, Anders, How do Intense Magnetic Storms Affect a Space Elevator?, IAC-13, paper and presentation, Beijing, Oct 2013.
- Kamogawa, Masashi Atmospheric electricity modulation caused by space elevator. IAC-18, paper and presentation, Bremen, Oct 2018.
- Knapman, J. (2005), "Dynamically Supported Launcher," Journal of the British Interplanetary Society, Vol. 58, No. 3/4, pp. 90-102
- Knapman, J. (2009), "The Space Cable: Capability and Stability," Journal of the British Interplanetary Society, Vol. 62, No.6, pp. 202-210
- Knapman, J. (2010), "Diverse Configurations of the Space Cable," 61st International Astronautical Congress, Prague, Czech Republic, 27 September 27-1 October 2010
- Knapman, J. (2011), "Space Elevator Stage I," 62nd International Astronautical Congress, Cape Town, South Africa, 3-7 October 2011
- Knapman, John, The Space Elevator in the Earth's Atmosphere, Via Ad Astra, Vol 1, 2015.
- Knapman, J. and Lofstrom, K. (2011), "Space Elevator Stage I: Through the Stratosphere," 2011 Space Elevator Conference, Redmond, WA, 12-14 August 2011
- Knapman, J. (2012), "Benefits and Development of a High Stage One," 63rdInternational Astronautical Congress, Naples Italy, 1-5 October 2012.
- Lang, D. D., "Approximating Aerodynamic Response of the Space Elevator to Lower Atmospheric Wind," Space Exploration 2005, SESI Conference Series, Vol. 1, 2005.
- Lofstrom, K. (1985), "The Launch Loop," AIAA Paper 85-1368, July 1985.

- Loftus, J. P. and Stansbery, E. G. (1993), "Protection of Space Assets by Collision Avoidance.", 44th Congress of the International Astronautical Federation, Austria. IAA 6.4-93-752
- NASA. (2010), "Debris density charts from NASA Orbital Debris Program Office", May 2010.
- Pearson, Jerome, Eugene Levin, John Oldson, Joseph Carroll, "EDDE: ElectroDynamic Debris Eliminator for Active Debris Removal,", International Conference on Debris Removal, Chantilly VA, 8-10 December 2009
- Pearson, Jerome, Eugene Levin, John Oldson, and Joe Carroll, "EDDE: ElectroDynamic Debris Eliminator for Safe Space Operations," 13th Annual FAA/AIAA Commercial Space Transportation Conference, Arlington, VA, 10-11 February 2010.
- Pearson, J. / Levin, E. / Carroll, J., ENHANCING SPACE ELEVATOR SAFETY BY ACTIVE DEBRIS REMOVAL, CLIMB, Vol. I, 2011.
- Penny, Robert, Space Debris & Space Elevator, IAC-08, paper and presentation, Glasgow, Oct 2008.
- Penny, Robert, Space Elevator Debris Mitigation Policy, 2009 IAC, paper and presentation, Daejeon, Oct 2009.
- Penny Robert and Jones, Richard, "A Model for Evaluation of Satellite Population Management Alternatives," AFIT Master's Thesis, 1983.
- Position Paper on Orbital Debris. International Academy of Astronautics, Paris 2000.
- Penny, R., Swan, C. and Swan, P. (2011), "Space Elevator Survivability; Space Debris Mitigation". ISEC Position Paper #2010-1, International Space Elevator Consortium, Lulu, 2011.
- Rugescu, Radu, Debris Hazards Mitigation and Retrieval for Space Elevators, IAC-06, paper and presentation, Valencia, Oct 2006
- Swan, Peter, Robert Penny & Cathy Swan, "Space Elevator Survivability: Space Debris Mitigation," ISEC Position Paper #2010-1, International Space Elevator Consortium, Fall, 2010.
- Swan, Peter and Cathy Swan, Space Elevator Systems Architecture, Lulu.com publishers, 2007.
- Swan, P., Raitt, Swan, Penny, Knapman. International Academy of Astronautics Study Report, Space Elevators: An Assessment of the Technological Feasibility and the Way Forward, Virginia Edition Publishing Company, 2013.
- Swan, Peter, Robert "Skip" Penny, and Cathy Swan, Space Elevator Survivability – Space Debris Mitigation, Lulu.com, 2011.
- Weeden, Billiards in Space, The Space Review, Feb 23, 2009. www.thespacereview.com/article/1314.

- Ambartsumian, S.A. / Belubekyan, M.V. / Ghazaryan, K.B. / Ghazaryan, R. A, TRANSVERSE VIBRATION OF THE SPACE ELEVATOR TETHER WITH SPACEPORTS, CLIMB, Vol. II, 2013.
- Aslanov, Vladimir, MOTION OF THE SPACE ELEVATOR AFTER THE RIBBON RUPTURE, IAC-12, paper and presentation, Naples, Oct 2011.
- Beletsky, V.V., and E.M. Levin, "Dynamics of Space Tether Systems," Vol. 83 of Advances in the Astronautical Sciences, Univelt, San Diego, 1993 (English version).
- Benaroya, Haym, Space Elevator Cable Dynamics, IAC-06, paper and presentation, Valencia, Oct 2006.
- Cohen, S. S., Dynamics of a Space Elevator, Master of Engineering Thesis, McGill University, Montreal, Quebec, 2006.
- Cohen, S. S. and Misra, A. K., "Elastic Oscillations of the Space Elevator Ribbon," Journal of Guidance Control and Dynamics, Vol. 30, No. 6, pp. 1711-1717, 2007.
- Cohen, Stephen, Effects of Climber Transit on the Space Elevator Dynamics, IAC-07, paper and presentation, Naples, Oct 2007.
- Cohen, S.S., and A.K. Misra, "The effect of climber transit on the space elevator dynamics," Acta Astronautica 64, 538 (2009).
- Cohen, S. S. and Misra, A. K., "The Effect of Climber Transit on the Space Elevator Dynamics," Acta Astronautica, Vol. 64, pp. 538-553, 2009.
- Cohen, Stephen, Static Deformation of Space Elevator Tether due to Climber, IAC-14, paper and presentation, Toronto, Oct 2014.
- Cohen, Stephen, Space Elevator Dynamic Response to Payload Release. IAC-19, paper and presentation, Washington D.C., Oct 2019
- Edwards, Bradley and Eric Westling, Space Elevator – A Revolutionary Earth-to-Space Transportation System, BC Edwards publishing, 2002.
- Edwards, Bradley, The Space Elevator, NIAC Phase I Study Report, 2000.
- Edwards, B. The Space Elevator, NIAC Phase II Final Report, 2003.
- Edwards, B.C., "Design and Deployment of a Space Elevator," Acta Astronautica 47, 735 (2000).
- Evensberget, Dag, Mechanics of the Space Elevator Including Deployment and Failure Modes, IAC-07, paper and presentation, Naples, Oct 2007.
- Fujii, H.A., Watanabe, T., Kusagaya, T. and Sato, D., "Dynamics of Flexible Space Tether Equipped with a Crawler Mass," Journal of Guidance, Control, and Dynamics, Vol. 31, No.2, pp.436-440, March/April 2008.
- Fujii, Hironori, Dynamics of Space elevator in Response to Disturbances, IAC-13, paper and presentation, Beijing, Oct 2013.
- Hodges, D.H. and E.H. Dowell, "Nonlinear equations of motion for the elastic bending and torsion of twisted nonuniform rotor blades," NASA Technical Note D-7818 (1974).
- Inoue, Fumihiro, Dynamic Behavior and Mechanism of Driving Roller for Climber Model in Space Elevator, IAC-16, paper and presentation, Guadalajara, Oct 2016
- Ishige, Yuuki & Satorni Kawamoto. "Study on Electrodynamic Tether System for Space Debris Removal." (IAF-02-A.7.04.) 53rd International Astronautical Congress, 2002.

- Ishikawa, Yoji, The Space Elevator Construction Concept, Obayashi Corporation, 2013, IAC-13-D4.3.6.
- Ishikawa, Yoji, Impact of Ascending and Descending Climbers on Space Elevator Cable Dynamics, IAC-16, paper and presentation, Guadalajara, Oct 2016.
- Ishikawa, Yoji, Space Elevator Cable's Oscillation Caused in Space Thermal Environment. IAC-19, paper and presentation, Washington D.C., Oct 2019
- Jorgensen, Anders, Space Elevator Interaction with the Space Environment: Numerical Simulations, IAC-05, paper and presentation, Fukuoka, Oct 2005.
- Jorgensen, Anders, Dynamics of the proposed space elevator under the influence of magnetospheric electric and magnetic fields, IAC-06, paper and presentation, Valencia, Oct 2006.
- Jorgensen, Anders, The Interaction of a Conducting Space Elevator with Magnetic and Electric Fields in the Near-Earth Space Plasma, IAC-12, paper and presentation, Naples, Oct 2011.
- Jorgensen, A.M. and S.E. Patamia, "How Do Intense Magnetic Storms Affect a Space Elevator?" 64th International Astronautical Conference, Beijing, IAC-13-D4.3, 8X18785 (2013).
- Knapman, John, Stability of the Space Cable, IAC-06, paper and presentation, Valencia, Oct 2006
- Knapman, John, "Benefits and Development of High Stage One for the Space Elevator," Naples IAC presentation and paper, IAC-12 D4.6.
- Knapman, John, Improving Stability of the Space Cable, IAC-08, paper and presentation, Glasgow, Oct 2008.
- Knapman, John, "A Multi-Stage Elevator," Paper given at 2016 ISEC Space Elevator Conference, Seattle, 19-21 August 2016.
- Knapman, John, Maintaining stability of the multi-stage space elevator, IAC-18, paper and presentation, Bremen, Oct 2018.
- Keshmiri, Mehdi, Consideration of Tether Elasticity in the Deployment Phase of a Space Elevator System, IAC-13, paper and presentation, Beijing, Oct 2013.
- Lang, D. D., "Space Elevator Dynamic Response to In-Transit Climbers," 1st International Conference on Science, Engineering, and Habitation in Space, Albuquerque, NM, Space Engineering and Science Inst., Paper 10152148, 2006.
- Lang, David, Space Elevator Dynamic Response to In-Transit Climbers, Via Ad Astra, Vol 1, 2015.
- Lades, Martin, "Climber-Tether Interface for a Space Elevator, ISEC Conference, Seattle, 2013.
- Lenard, Roger, Mid-Earth Momentum Transfer Tether, IAC-08, paper and presentation, Glasgow, Oct 2008.
- Li, Gangqiang, Dynamics of Partial Space Elevator with Parallel Tethers and Multiple Climbers, IAC-19, paper and presentation, Washington D.C., Oct 2019
- Lorenzini, C. and Cosmo, M., "Wave Propagation in the Tether Elevator/Crawler System," Acta Astronautica, Vol. 21, No. 8, pp. 545-552, 1990.
- Mazzoleni, Andre, Deployment Dynamics of Space Elevator Ribbon, IAC-11, paper and presentation, Cape Town, Oct 2011.
- McInnes, Colin, Novel Payload Dynamics on Space Elevator Systems, IAC-05, paper and presentation, Fukuoka, Oct 2005.
- Misra, Arun, Towing of Space Debris Using a Tether, IAC-17, paper and presentation, Adelaide, Australia, Sept 2017.

- Modi, V. J., Bachman, S., and Misra, A. K., "Dynamics and Control of a Space Station Based Tethered Elevator System," Acta Astronautica, Vol. 29, No. 6, pp. 429-449, 1993.
- Murakami, Daichi, Design and development of the tether moving system using nanosatellite, IAC-18, paper and presentation, Bremen, Oct 2018.
- Ohkawa, R., Uchiyama, K., and Fujii, H. A., "The Effect of Disturbance on Space Elevator Dynamics with Flexibility," 61th International Astronautical Congress, Prague, IAC-10-D4. 4. 5, 27 Sep. -1 Oct. 2010.
- Patamia, Steven, Analytic Model of Dynamic Response of Proposed Space Elevator to Anchor Point Repositioning, IAC-05, paper and presentation, Fukuoka, Oct 2005.
- Patamia, Steven, Approaches to taming oscillations of terrestrial space elevators and reducing their exposure to van Allen radiation., IAC-14, paper and presentation, Toronto, Oct 2014.
- Pearson, J. "The Orbital Tower: A Spacecraft Launcher Using the Earth's Rotational Energy," Acta Astronautica 2, 785 (1975).
- Penny, Robert. Swan, Peter, & Cathy Swan, "Space Elevator Concept of Operations," ISEC Position Paper #2012-1, International Space Elevator Consortium, Fall, 2013.
- Perek, Lubos, Space Elevator: Stability, IAC-06, paper and presentation, Valencia, Oct 2006.
- Rugescu, Radu, Loads during Anchoring Dynamics from Earth Orbit, IAC-06, paper and presentation, Valencia, Oct 2006.
- Rugescu, Radu, Independent dynamics and stability of twin tethered objects, 2009 IAC, paper and presentation, Daejeon, Oct 2009.
- Rugescu, Radu, Soft Landing Dynamics Study with Extension to Elevator Anchoring, IAC-08, paper and presentation, Glasgow, Oct 2008.
- Rugescu, Radu, An Inverse Dynamics Method for Soft Landing and Anchoring Planning, IAC-07, paper and presentation, Naples, Oct 2007.
- Sidi, M., Spacecraft Dynamics and Control: A Practical Engineering Approach, Cambridge University Press, pp. 28-62, 1997.
- Srinil, N., G. Rega and S. Chucheepsakul, "Three-dimensional non-linear coupling and dynamic tension in the large amplitude free vibrations of arbitrarily sagged cables," Journal of Sound and Vibration 269, 823 (2004).
- Swan, P., Raitt, Swan, Penny, Knapman. International Academy of Astronautics Study Report, Space Elevators: An Assessment of the Technological Feasibility and the Way Forward, Virginia Edition Publishing Company, 2013.
- Swan, Peter, Robert "Skip" Penny, and Cathy Swan, Space Elevator Survivability – Space Debris Mitigation, Lulu.com, 2011.
- Takeichi, N., "Geostationary stationkeeping control of a space elevator during initial cable deployment", 61st International Astronautical Congress, Prague, Czech Republic, October 2010, paper No. IAC-10-D.4.4.7.
- Takeichi, N., "Geostationary station keeping control of a space elevator during initial cable deployment", Acta Astronautica, Vol. 70, pp. 85-94, 2012.
- Troger, Hans, On the Stability of the Track of the Space Elevator, IAC-06, paper and presentation, Valencia, Oct 2006
- Uchiyama, K., Iijima, K., and Fujii, H. A., "Construction of Space Elevator Model Using Absolute Nodal Coordinate," Transactions on Advanced Research IPSI Bgd Internet Research Society, ISSN 1820-4511, Vol. 9, No. 2, pp.8-12, July 2013.

- West, Icole, Small Scale SE Ribbon Dynamics: Finite Element Analyses of Regional Phenomena, IAC-04, paper and presentation, Vancouver, Oct 2004.
- Williams, P., "Dynamic Multibody Modeling for Tethered Space Elevators," Acta Astronautica, Vol. 65, No. 3-4, pp.399-422, Aug-Sept. 2009.
- Williams, P. and Ockels, W., "Climber motion optimization for the tethered space elevator, Acta Astronautica, doi:10.1016/ j. actaastro.2009.11.003, 2009.
- Williams, Paul, Dynamic Multibody Modeling for Tethered Space Elevators, IAC-07, paper and presentation, Naples, Oct 2007.
- Wright, Dennis, "Space Elevator Research and ISEC Studies," Presentation at the National Space Society Conference, St. Louis, May 2017.
- Wright, Dennis, A Hardware Space Elevator Simulator, International Space Elevator Conference, paper and presentation, Seattle, Aug 2018.
- Woo,P. and Misra, A.K. "Dynamics of a partial space elevator with multiple climbers," Acta Astronautica, doi:10.1016/j.actaastro.2010.04.023, 2010.
- Woo, Pamela, Dynamics of a Partial Elevator with Multiple Climbers, IAC-08, paper and presentation, Glasgow, Oct 2008.
- Wright, Dennis, S. Avery, J. Knapman, M. Lades, P. Roubekas, P. Swan; Design Considerations for a Software Space Elevator Simulator, ISEC Study Report, lulu.com, 2017
- Yamagiwa, Yoshiki, Study about the Performance for Simultaneous Deployment of the Cables from GEO Station under the Space Elevator Construction, IAC-16, paper and presentation, Guadalajara, Oct 2016
- Yamagiwa, Yoshiki, Optimum Control of Cable Deployment of Space Elevator from GEO station in two directions, IAC-18, paper and presentation, Bremen, Oct 2018.
- Yasaka, Tetsuo, Dynamics and Stability of Space Elevator during Initial Deployment, 2009 IAC, paper and presentation, Daejeon, Oct 2009.
- Yasaka, Tetsuo, Orbital Motion of VEry Long Systems, IAC-18, paper and presentation, Bremen, Oct 2018.
- Yokochi, Masanori, Experimental Study on Effect of Climbing Rider on Lateral Deviation of Space Elevator, IAC-14, paper and presentation, Toronto, Oct 2014.
- Yokota, Shun, The effects of payload transportation on the tethered ssystems in low earth orbit, IAC-18, paper and presentation, Bremen, Oct 2018.

Electrodynamics

- Allison, J. et al., "Recent Developments in Geant4," Nuclear Instruments and Methods in Physics Research A, 186 (2016).
- Fujii, H.A. et al. (2009). Sounding rocket experiment of bare electrodynamic tether system. Acta Astronautica, vol. 64, p.313-324.
- Gilchrist, B. et. al. (1998). Enhanced electrodynamic tether currents due to electron emission from a neutral gas discharge: Results from the TSS-1R mission. Geophysical Research Letters, Vol. 25, No. 4, pp. 437-440, February 15, 1998.
- Ginet, G.P. et al., "AE9, AP9 and SPM: New Models for Specifying the Trapped Energetic Particle and Space Plasma Environment," Space Science Reviews 179, 579 (2013).
- Jorgensen, A.M., S.E. Patamia and B. Gassend, "Passive radiation shielding considerations for proposed space elevator," Acta Astronautica 60, 198 (2007).
- Jorgensen, Anders, How do Realistic Magnetospheric Fields Affect Space Elevators?, IAC-16, paper and presentation, Guadalajara, Oct 2016
- Tsyganenko, N.A., "A Magnetospheric Magnetic Field Model with a Warped Tail Current Sheet," Planetary and Space Science 37, 5 (1989).
- Wright, D. "Electric Currents on the Space Elevator," International Space Elevator Conference, Seattle, August 2013.

Earth Port

- Edwards, Bradley and Eric Westling, Space Elevator – A Revolutionary Earth-to-Space Transportation System, BC Edwards publishing, 2002.
- Edwards, Bradley, The Space Elevator, NIAC Phase I Study Report, 2000.
- Edwards, B. The Space Elevator, NIAC Phase II Final Report, 2003.
- Fitzgerald, Michael, "Space Elevator Sequences and Initial Operational Capability," Paper given at 2016 ISEC Space Elevator Conference, Seattle, 19-21 August 2016.
- Fukazawa, Takeyuki, A Study of Marine Node in Construction stage of the space elevator, IAC-18, paper and presentation, Bremen, Oct 2018.
- Hall, Vern, R. Penny, P. Glaskowsky, S. Schaeffer, Design Considerations for Space Elevator Earth Port, ISEC Study Report, www.lulu.com, 2016
- Hall, Vernon, "Earth Port Access City: The Case for Hawaii," presented at 2017 ISEC Conference, Seattle, 25-27 Aug 2017.
- Ishikawa, Yoji, The Space Elevator Construction Concept, Obayashi Corporation, 2013, IAC-13-D4.3.6.
- Knapman, John, P. Glaskowsky, D. Gleeson, V. Hall, D. Wright, M. Fitzgerald, P. Swan, Design Considerations for the Multi-Stage Space Elevator, ISEC Study Report, lulu.com, 2018.
- Penny, Robert, Concept for a Space Elevator Earth Port, IAC-16, paper and presentation, Guadalajara, Oct 2016
- Swan, P., Raitt, Swan, Penny, Knapman. International Academy of Astronautics Study Report, Space Elevators: An Assessment of the Technological Feasibility and the Way Forward, Virginia Edition Publishing Company, 2013.
- Swan, Peter, Space Elevator Design Aspects for the Environment, IAC-12, paper and presentation, Naples, Oct 2011.
- Swan, Peter, Characteristics of Space Elevator Apex Anchor, International Space Elevator Conference, paper and presentation, Seattle, Aug 2018.
- Takeyuki Fukazawa, Study of Tension Control Components on Earth surface platform for Space Elevator System. IAC-19, paper and presentation, Washington D.C., Oct 2019

Multi-Stage

- Knapman, John, Space Elevator in the Atmosphere, IAC-14, paper and presentation, Toronto, Oct 2014.
- Knapman, John, "Benefits and Development of High Stage One for the Space Elevator," Naples IAC presentation and paper, IAC-12 D4.6.
- Knapman, John, "A Multi-Stage Elevator," Paper given at 2016 ISEC Space Elevator Conference, Seattle, 19-21 August 2016.
- Knapman, John, Space Elevator Stage I, IAC-11, paper and presentation, Cape Town, Oct 2011.
- Knapman, John, Benefits and Development of a High Stage One for the Space Elevator, IAC-12, paper and presentation, Naples, Oct 2011.
- Knapman, John, The Space Elevator in the Earth's Atmosphere, Via Ad Astra, Vol 1, 2015.
- Knapman, John, The Space Elevator Tower, IAC-16, paper and presentation, Guadalajara, Oct 2016
- Knapman, John, "Stability of the Multi-stage Space Elevator," presented at 2017 ISEC Conference, Seattle, 25-27 Aug 2017.
- Knapman, John, P. Glaskowsky, D. Gleeson, V. Hall, D. Wright, M. Fitzgerald, P. Swan, Design Considerations for the Multi-Stage Space Elevator, ISEC Study Report, lulu.com, 2018.
- Knapman, John, The Multi-Stage Elevator: 2019 Update. presented at 2019 International Space Elevator Conference, Seattle, 16-18 Aug 2019.
- Kanpman, John, Progress Report on the Multi-stage Space Elevator. IAC-19, paper and presentation, Washington D.C., Oct 2019

- Edwards, Bradley and Eric Westling, Space Elevator – A Revolutionary Earth-to-Space Transportation System, BC Edwards publishing, 2002.
- Fitzgerald, M, R. Penny, P. Swan, C. Swan, Space Elevator Architectures and Roadmaps, ISEC Study Report, lulu.com, 2015
- Fitzgerald, Michael, "Space Elevator Sequences and Initial Operational Capability," Paper given at 2016 ISEC Space Elevator Conference, Seattle, 19-21 August 2016.
- Hall, Vern, R. Penny, P. Glaskowsky, S. Schaeffer, Design Considerations for Space Elevator Earth Port, ISEC Study Report, www.lulu.com, 2016
- Ishikawa, Yoji, The Space Elevator Construction Concept, Obayashi Corporation, 2013, IAC-13-D4.3.6.
- Penny, Robert. Swan, Peter, & Cathy Swan, "Space Elevator Concept of Operations," ISEC Position Paper #2012-1, International Space Elevator Consortium, Fall, 2013.
- Penny, Robert, Design Considerations for Geo Node, Apex Anchor and Communications Architecture ISEC Study underway 2017.
- Swan, P., Raitt, Swan, Penny, Knapman. International Academy of Astronautics Study Report, Space Elevators: An Assessment of the Technological Feasibility and the Way Forward, Virginia Edition Publishing Company, 2013.

95

- Angel, R. (2006), "Feasibility of Cooling the Earth with a Cloud of Small Spacecraft near L1," Proceedings of the National Academy of Sciences, v 103, n46, 2006 November 14, 2006. Pp. 17184–17189. http://www.ncbi.nlm.nih.gov/pmc/articles/PMC1859907
- Aoki, Yoshio Jun Maeda, Linear Direct Drive Motor Mechanism for Use in tethered satellites, IAC-18, paper and presentation, Bremen, Oct 2018.
- Bartoszek, Larry, "Getting the Mass of the First Construction Climber Under 900 Kg," ISEC Conference, Seattle, 2013.
- Bou, Elisenda,Laser and the Space Elevator: an Approachment, IAC-07, paper and presentation, Naples, Oct 2007.
- Edwards, Bradley and Eric Westling, Space Elevator – A Revolutionary Earth-to-Space Transportation System, BC Edwards publishing, 2002.
- Edwards, Bradley, Results from the First Annual Space Elevator Climber Competition, IAC-05, paper and presentation, Fukuoka, Oct 2005.
- Hein, Andreas, analysis of possible changes in spacecraft design due to the usage of a space elevator for transportation, 2009 IAC, paper and presentation, Daejeon,
- Inoue, Fumihiro, Experiment study of Climber Mechanism with Cross Roller System for Heavy Load in Space Elevator, IAC-18, paper and presentation, Bremen, Oct 2018.
- Ishikawa, Yoji, The Space Elevator Construction Concept, Obayashi Corporation, 2013, IAC-13-D4.3.6.
- Hinton, G., HEAT DISSIPATION ISSUES IN SPACE ELEVATOR CLIMBERS, CLIMB, Vol. I, 2011.
- Ishikawa, Yoji, Inoue, Fumihiro, Experimental Study on Climber Mechanism Applying Cross Roller System for Small Manned Space Elevator. IAC-19, paper and presentation, Washington D.C., Oct 2019
- Inoue, Fumihiro, Development and Driving Experiment of Climber Mechanism for Heavy Load in Space Elevator, IAC-17, paper and presentation, Adelaide, Australia, Sept 2017.
- Kakuta, Tomohiro, Thermal Study for the STARS-E Climber's Mission, IAC-17, paper and presentation, Adelaide, Australia, Sept 2017.
- Keshmiri, Mehdi, On the deployment of a subsatellite in a space elevator system, IAC-12, paper and presentation, Naples, Oct 2011.
- Knapman, John, Diverse Configurations of the Space Cable, IAC presentation and paper, IAC-10 Session D4.
- Knapman, John, "Tether Climber at Constant Power," ISEC Conference, Seattle, 2013.
- Knapman, John, "Benefits and Development of High Stage One for the Space Elevator," Naples IAC presentation and paper, IAC-12 D4.6.
- Lades, Martin, "Climber-Tether Interface for a Space Elevator, ISEC Conference, Seattle, 2013.
- Lades, Martin, Wireless Power Transfer to a Moving Vehicle: Explorations with the Kansas City team for the NASA/Spaceward power beaming challenge, IAC presentation and paper, IAC-10 Session D4.
- Laine, Michael, Chapter 3: Spacecraft at: http://www.mill-creek-systems.com/HighLift/chapter3.html
- Lofstrom, Keith, "Acoustic Powered Tether Climbers," presented at 2017 ISEC

Conference, Seattle, 25-27 Aug 2017.

- Mankins, J. (2011), "Space Solar Power, The First International Assessment of Space Solar Power: Opportunities, Issues and Potential Pathways Forward", IAA, October 2011.
- Nakashima, Kenji, Design of reel-type tether deployment mechanism and analysis of tether deployment dynamics in the microsatellite STARS-E for verifying the basic technology of space elevator, IAC-17, paper and presentation, Adelaide, Australia, Sept 2017.
- Ohno, Shuichi, Japanese Space Elevator Competitions and Challenges, Journal of British Interplanetary Society, Vol 69, No 06/07, Dec 2016.
- Ohno, Shuichi, "Space Mechatronics - A Global Challenge of SE Communities," presented at 2017 ISEC Conference, Seattle, 25-27 Aug 2017.
- Pasko, Vadym, An assessment of the technological feasibility of applying GEO-based solar pumped lasers for feeding the Space Elevator exoatmospheric climber, IAC-17, paper and presentation, Adelaide, Australia, Sept 2017.
- Penny, Robert "Skip", "Tether Climber Operational Phases," ISEC Conference, Seattle, 2013.
- Penny, Robert, Space Elevator Climber Operations, IAC-14, paper and presentation, Toronto, Oct 2014.
- Penny, R., P. Swan, C. Swan, J. Knapman, P. Glaskowsky, Design Considerations for Space Elevator Tether Climbers, ISEC Study Report, www.lulu.com, 2014
- Purang, Deepak (n.d.), "Space sunshade may one day reduce global warming." Editorial. http://www.streetdirectory.com/travel_guide/14921/gadgets/space_sunshade_may_one_day_reduce_global_warming.html
- Robinson, Peter, Space Elevator Climber : Tether Interface, Reliability and Other Considerations. presented at 2019 International Space Elevator Conference, Seattle, 16-18 Aug 2019.
- Robinson, Peter, Tether Material and the Climber Interface. Presented at NSS International Space Development Conference, Washington, D.C. June 7-9 June, 2019.
- Semon, Ted, "Hybrid Tether Climber," ISEC Conference, Seattle, 2013.
- Shelef, B., "A Solar-Based Space Elevator Architecture," Spaceward Foundation, 2008. http://www.spaceward.org/elevator-library#SW
- Shelef, B., "Segment Based Ribbon Architecture"., In Proc. of 3rd International Space Elevator Conference, June 2004.
- Shelef, B., "The Space Elevator Feasibility Condition", Climb Journal, Volume 1, Number 1, p. 87. And in - Spaceward Foundation, 2008. Available at: http://www.spaceward.org/elevator-library#SW
- Shelef, B., "Space Elevator Power System Analysis and Optimization, Spaceward Foundation, 2008. Available at: http://www.spaceward.org/elevator-library#SW
- Smith, Frederick G., Accetta, Joseph S., and Shumaker, David L., The Infrared and Electro-Optical Systems Handbook: Atmospheric Propagation of Radiation, ANN ARBOR MI, 1993.
- Suemori, K., "Film-shaped thermoelectric conversion elements can be produced in print" Available at: http://www.aist.go.jp/aist_j/aistinfo/aist_today/vol12_04/p17.html
- Swan, P., Raitt, Swan, Penny, Knapman. International Academy of Astronautics Study Report, Space Elevators: An Assessment of the Technological Feasibility and the Way Forward, Virginia Edition Publishing Company, 2013.

- Swan, Peter, What if? Space Solar Power was Enabled by Space Elevators, IAC-08, paper and presentation, Glasgow, Oct 2008.
- Swan, Peter, Space Elevator Tether Climbers – Normal Spacecraft?, IAC-14, paper and presentation, Toronto, Oct 2014.
- Swan, P., NASA's Space Elevator Games: A History, Journal of British Interplanetary Society, Vol 69, No 06/07, Dec 2016.
- TSM , "Technological Strategy Zmap 2010 – Energy", Ministry of Economy, Trade and Industry. Available at: http://www.meti.go.jp/policy/economy/gijutsu_kakushin/kenkyu_kaihatu/str2010 download.html
- Tsuchida A. et al, "New Space Transportation System-Space Train (Elevator): World trends and Japanese Space Train Concept", Technical report of IEICE. SANE 109(101), 93-98, 2009-06-18.
- Yoshino, Kazuyoshi, Experimental Study on Speed Control of Rider on Twisted Tape Tether Using Image Processing, IAC-13, paper and presentation, Beijing, Oct 2013.
- Yokota, Shun, Regarding the Effect of a Climber's Motion on the Tethered Satellite System, IAC-17, paper and presentation, Adelaide, Australia, Sept 2017.

GEO Node

- Edwards, Bradley and Eric Westling, Space Elevator – A Revolutionary Earth-to-Space Transportation System, BC Edwards publishing, 2002.
- Fitzgerald, Michael, Vern Hall, Peter Swan, Cathy Swan, "Design Considerations for Space Elevator Apex Anchor and GEO Node," ISEC Study Report, 2017.
- Fitzgerald, Michael, "Space Elevator Sequences and Initial Operational Capability," Paper given at 2016 ISEC Space Elevator Conference, Seattle, 19-21 August 2016.
- Ishikawa, Yoji, The Space Elevator Construction Concept, Obayashi Corporation, 2013, IAC-13-D4.3.6.
- Ishikawa, Yoji, Critical Technologies for Space Elevator's GEO Nodes, Earth Port, Gates and Communications, IAC-17, paper and presentation, Adelaide, Australia, Sept 2017.
- Swan, P., Raitt, Swan, Penny, Knapman. International Academy of Astronautics Study Report, Space Elevators: An Assessment of the Technological Feasibility and the Way Forward, Virginia Edition Publishing Company, 2013.
- Swan, Peter, Michael Fitzgerald, "Space Elevator GEO Node and Apex Anchor Architectures," IAC-17, paper and presentation, Adelaide, Australia, Sept 2017.

- Edwards, Bradley and Eric Westling, Space Elevator – A Revolutionary Earth-to-Space Transportation System, BC Edwards publishing, 2002.
- Fitzgerald, Michael, "Space Elevator Sequences and Initial Operational Capability," Paper given at 2016 ISEC Space Elevator Conference, Seattle, 19-21 August 2016.
- Fitzgerald, M, R. Penny, P. Swan, C. Swan, Space Elevator Architectures and Roadmaps, ISEC Study Report, lulu.com, 2015.
- Fitzgerald, Michael, Vern Hall, Peter Swan, Cathy Swan, "Design Considerations for Space Elevator Apex Anchor and GEO Node," ISEC Study Report, 2017.
- Ishikawa, Yoji, The Space Elevator Construction Concept, Obayashi Corporation, 2013, IAC-13-D4.3.6.
- Swan, P., Raitt, Swan, Penny, Knapman. International Academy of Astronautics Study Report, Space Elevators: An Assessment of the Technological Feasibility and the Way Forward, Virginia Edition Publishing Company, 2013.
- Swan, Peter, Michael Fitzgerald, "Space Elevator GEO Node and Apex Anchor Architectures," IAC-17, paper and presentation, Adelaide, Australia, Sept 2017.

Asteroids, Lunar and Mars Elevators

- Bezrodny, G., N. Greenfeld, A. Tatievsky, R. Qedar, O. Reuven, A. Kogan, "Lunar Space Elevator, "Jacob's Ladder," Aerospace Faculty Technion, Haifa, Israel, 2009.
- Eubanks, T.M., A space elevator for the far side of the moon, Annual Meeting of the Lunar Exploration Analysis Group, LPI Contributions (2013), p. 7047
- Eubanks, T.M., C.Maccone,C.F.Radley Lunar farside radio astronomy base facilitated by lunar elevator Annual Meeting of the Lunar Exploration Analysis Group, Vol.1863 of LPI Contributions (2015), p. 2014
- Ganapathy, Rohan, Conceptual Colonization of Space Using Space-Elevators from Mars' natural Satellite "Phobos", IAC-13, paper and presentation, Beijing, Oct 2013.
- Guerman, Anna, DYNAMICS OF A PLANET-TETHERED SPACECRAFT, IAC-11, paper and presentation, Cape Town, Oct 2011.
- Guerman, Anna, Dynamics of moon elevator, IAC-14, paper and presentation, Toronto, Oct 2014.
- Guerman, Anna, Space Elevator operation in proximity of asteroids, IAC-19, paper and presentation, Washington D.C., Oct 2019
- Heckman, Fuller-Clarke Sphere, International Space Elevator Conference, paper and presentation, Seattle, Aug 2018.
- Lades, Martin and Lake Matthew Team, "Omaha Trail to Lake Matthew," presented at 2017 ISEC Conference, Seattle, 25-27 Aug 2017.
- Lades, Martin, Mars Lift UpdateIAC-18, paper and presentation, Bremen, Oct 2018.
- Laine, Michael, "Leveraging the lunar Space Elevator for L-1 Development," Presentation at the National Space Society Conference, St. Louis, May 2017.
- Mistry, Ashish, ELEVATOR TRANSPORTATION BETWEEN MARS & IT'S MOONS, IAC-08, paper and presentation, Glasgow, Oct 2008.
- Moravec, H., A non-synchronous orbital skyhook J. Astronautical Sci., 25 (1977), pp. 307–322
- Parkinson, Robert, Partial Beanstalks for Mars Exploration, IAC-04, paper and presentation, Vancouver, Oct 2004.
- Pearson, Jerome, The Lunar Space Elevator, IAC-04, paper and presentation, Vancouver, Oct 2004
- Pearson, J., E. Levin, J. Oldson, and H. Wykes, Lunar Space Elevators for CISLUNAR Space Development, NIAC Phase I Final Technical Report, 2 May 2005.
- Pearson, J., Eugene Levin, John Oldson, and Harry Wykes, "The Lunar Space Elevator," Space Technology, Vol. 25, No. 3-4, pp. 203-209, 2005.
- Pearson, J., Eugene Levin, John Oldson, and Harry Wykes, "Lunar Space Elevators for Cis-Lunar Transportation," International Conference, Moon Base: A Challenge for Humanity, Venice Workshop, Venice, Italy, 26-27 May 2005.
- Pearson, j. "Space Elevators for Earth and Moon," presented at the International Space Development Conference, ISDC 2007, Dallas, TX, 24-28 May 2007.
- Pearson, Jerome, LUNAR ANCHORED SATELLITE TEST, CLIMB, Vol. II, 2013.

- Pearson, Jerome, Anchored Lunar Satellites for Cislunar Transportation and Communication, Via Ad Astra, Vol 1, 2015.
- Radley, Charles, Lunar Skylift: Cable Oscillations and their Treatment, IAC-14, paper and presentation, Toronto, Oct 2014.
- Radley, Charles, Lunar Elevator - Payload transfer on Earthbound flow, IAC-14, paper and presentation, Toronto, Oct 2014.
- Razzaghi, Kaveh, Conceptual Design and Technology Roadmap for a Lunar Space Elevator, IAC-17, paper and presentation, Adelaide, Australia, Sept 2017.
- Swan, Peter, First Space Elevator: on the Moon, Mars or the Earth?, IAC presentation and paper, IAC-10 Session D4.
- Swan, Peter, Opening up earth-moon enterprise with a space elevator, New Space, 3 (2015), pp. 213–217
- Wang, Xiaohui, Dynamics Research of Initial Tether Deployment of Lunar Space Elevator, IAC-17, paper and presentation, Adelaide, Australia, Sept 2017.

Miscellaneous

- Ambrose, S. (2000), "Nothing like it in the World". Simon & Schuster, New York, 2000.
- Clarke, A. C. (2003), "Discussion by GEO satellite relay from Sri Lanka, 2nd Annual International Space Elevator Conference, 2003, Sante Fe, NM.
- Dwyer, J. R. and Smith, D. M. (2012), "Deadly Rays from Clouds," Scientific American, 307, August 2012. pp. 54-59.
- EUSPEC (2011), "Evaluation", Available at: http://euspec.warr.de/handbook
- Highlift at: http://www.mill-creek-systems.com/HighLift/chapter3.html
- Gale, S. F. (2011), "Biggest isn't always better," PM Network, March 2011. Available at: http://www.pegasus-global.com/newsletters/201104/Patricia_D_Galloway_Columbia_River_Crossing_Leadership_On_Megaprojects.pdf
- Gelb, M. J. (2012), http://thinkexist.com/quotes/michael_j._gelb, June 2012.
- Johnson, L. et al (2003). Propulsive Small Expendable Deployer System (ProSEDS) Experiment Mission Overview and Status. In 39th AIAA ASME SAE ASEE Joint Propulsion Conference.
- Levin, E.M. (2007). Dynamic Analysis of Space Tether Missions. Advances in the Astronautical Sciences, vol. 126.
- McCoy, J.E. et al. (1995). Plasma Motor-Generator (PMG) Flight Experiment Results. In the fourth Conference On Tethers in Space, pp. 57-82.
- Meulenberg, A. and Karthik Balaji P.S., "The LEO Archipelago: A System of Earth-Rings for Communications, Mass-Transport to Space, Solar Power, and Control of Global Warming," Acta Astronautica 68 (2011), iss. 11-12 Jun 2011 pp. 1931-1946, doi: 10.1016/j.actaastro.2010.12.002 arXiv:1009.4043v1.
- Meulenberg, A. and Wan, T., C., "LEO-Ring-Based Communications Network," Proceedings of Space, Propulsion & Energy Sciences International Forum (SPESIF-11, March 15-17, 2011), Physics Procedia, Volume 20, 2011, Pages 232-241, edited by Glen A. Robertson.
- Meulenberg, A. and Poston, T., "Sling-on-a-Ring: Structural elements for a Space Elevator to LEO," Proc. of SPESIF-11, March 15-17, 2011, Physics Procedia Volume 20, 2011, pp 222-231, Space, Propulsion & Energy Sciences International Forum edited by Glen A. Robertson.
- Gaurav Sharma and Andrew Meulenberg, "Collosal Carbon Tubes as Tethers for a Space Elevator," Conf. Proceedings, International Space Elevator Consortium, Seattle, Washington, USA, 23 Aug. 2012
- Meulenberg, A, "Future Perspectives in Space Solar Power Generation, Transmission, and Storage,"Keynote speaker, International Multidisciplinary Conf. on Solar Energy (IMDCSE-2012), Chennai, India, 1-3 Feb. 2012
- Meulenberg, A., "Sling on a ring: mass- and man-transport to space," presented at the 62nd International Astronautical Congress 2011, Cape Town S.A. Oct. 3-7
- Meulenberg, A., T. C. Wan, "Fiber-optic, LEO-based, communications ring," presented at the 62nd International Astronautical Congress 2011, Cape Town S.A. Oct. 3-7
- Meulenberg, A., T. Poston, "Compressive members for a space elevator to LEO," presented at the 62nd Int. Astronautical Congress 2011, Cape Town S.A. Oct. 3-7

- R. Suresh and A. Meulenberg, "A LEO-based solar-shade system to mitigate global warming," Proc. of Int. Astronautics Congress, IAC-2009, Daejeon, Korea, Oct. 2009, and http://arxiv.org/abs/1504.05148
- Meulenberg, A., R. Suresh, S. Ramanathan, and Karthik Balaji P.S., "LEO-Based Space-Elevator Development using Available Materials and Technologies," IAC-2009 Daejeon, Korea, Oct. 2009
- Meulenberg, A., R. Suresh, S. Ramanathan, and Karthik Balaji P. S., "Solar Power from LEO," Proc. of International Conf. on Energy and Environment, 19-21 Mar. 2009, Chandigarh, India
- Meulenberg, A., R. Suresh, and S. Ramanathan, "LEO-Based Optical/Microwave Terrestrial Communications" presented International Astronautics Congress, IAC-2008, Glasgow, Scotland, arXiv:1009.5506v1
- Meulenberg, A., Karthik Balaji P. S, R. Suresh and S. Ramanathan, "Sling-On-A-Ring: A Realizable Space Elevator to LEO?" presented at IAC-2008, Glasgow, Scotland, http://www.iafastro.net/iac/archive/browse/IAC-08/D4/1/366/ and as Appendix in http://www.sciencedirect.com/science?_ob=ArticleURL&_udi=B6V1N-51YB077-1&_user=4187955&_coverDate=01%2F15%2F2011&_rdoc=1&_fmt=high&_orig=gateway&_origin=gateway&_sort=d&_docanchor=&view=c&_acct=C000012438&_version=1&_urlVersion=0&_userid=4187955&md5=11bd165886015eb911dd7758598d47dc&searchtype=a
- Pearson, J. "The Real History of the Space Elevator," paper IAC-06-D4.3.01, 57th International Astronautical Congress, Valencia, Spain, 2-6 October 2006.
- Shinobu Doi (2011), "JEM Extended Utilization for Exposed Experiments using JEM Airlock and Robotics", 15th Annual ISU International Symposium, Strasbourg, France, Feb 16, 2011
- Teal Group (2012), "World Space Systems Briefing". Teal Group, 2012.
- The British Interplanetary Society. (2012), "Is space commercialization a myth?," Spaceflight Magazine, v54, n6, June 2012, p. 206
- Tsuchida, Akira, "IAA Study – "The Road to the Space Elevator Era," presented at 2017 ISEC Conference, Seattle, 25-27 Aug 2017.
- Tsyganenko, N., Fortran programs Geopack-2008, http://geo.phys.spbu.ru/~tsyganenko/modeling.html
- Volland, H.J. Geophys. Res. 78, 171 (1973); & D.J. Stern, Geophys. Res. 80, 595 (1975).
- VSO (Visual Satellite Observer's Homepage). http://www.satobs.org/noss.html, last accessed November 2010.
- Weeden, B. (2009), Billiards in Space. The Space Review, Feb 23, 2009. Available at: www.thespacereview.com/article/1314/1
- Whitesides, G. (2004), "Panel Discussion," The Space Elevator 3rd Annual International Space Elevator Conference, 30 June 2004, Washington,D.C.
- Wiese, Tim, A Journey of Student Space Elevator Development, IAC-18, paper and presentation, Bremen, Oct 2018.

Appendix G: Description of ISEC

Who We Are
The International Space Elevator Consortium (ISEC) is composed of individuals and organizations from around the world who share a vision of humanity in space.

Our Vision
A world with inexpensive, safe, routine, and efficient access to space for the benefit of all mankind.

Our Mission
The ISEC promotes the development, construction and operation of a space elevator infrastructure as a revolutionary and efficient way to space for all humanity.

What We Do
- Provide technical leadership promoting development, construction, and operation of space elevator infrastructures.
- Become the "go to" organization for all things space elevator.
- Energize and stimulate the public and the space community to support a space elevator for low cost access to space.
- Stimulate science, technology, engineering, and mathematics (STEM) educational activities while supporting educational gatherings, meetings, workshops, classes, and other similar events to carry out this mission.

A Brief History of ISEC
The idea for an organization like ISEC had been discussed for years, but it wasn't until the Space Elevator Conference in Redmond, Washington, in July of 2008, that things became serious. Interest and enthusiasm for a space elevator had reached an all-time peak and, with Space Elevator conferences upcoming in both Europe and Japan, it was felt that this was the time to formalize an international organization. An initial set of directors and officers were elected and they immediately began the difficult task of unifying the disparate efforts of space elevator supporters worldwide.

ISEC's first Strategic Plan was adopted in January of 2010 and it is now the driving force behind ISEC's efforts. This Strategic Plan calls for adopting a yearly theme to focus ISEC activities. Because of our common goals and hopes for the future of mankind off-planet, ISEC became an Affiliate of the National Space Society in August of 2013. In addition, ISEC works closely with the Japanese Space Elevator Association.

Our Approach
ISEC's activities are pushing the concept of space elevators forward. These cross all disciplines and encourage people from around the world to participate. The following activities are being accomplished in parallel:
- Yearly conference – International space elevator conferences were initiated by Dr. Brad Edwards in the Seattle area in 2002. Follow-on conferences were in Santa Fe (2003), Washington DC (2004), Albuquerque (2005/6 –smaller sessions), and Seattle (2008 to the present). Each of these conferences had multiple discussions across the whole arena of space elevators with remarkable concepts and presentations.
- Yearlong technical studies – ISEC sponsors research into a focused topic each year to ensure progress in a discipline within the space elevator project. The first such study was conducted in 2010 to evaluate the threat of space debris. The products from these

studies are reports that are published to document progress in the development of space elevators. They can be downloaded at www.isec.org.

- International Cooperation – ISEC supports many activities around the globe to ensure that space elevators keep progressing towards a developmental program. International activities include coordinating with the two other major societies focusing on space elevators: the Japanese Space Elevator Association and EuroSpaceward. In addition, ISEC supports symposia and presentations at the International Academy of Astronautics and the International Astronautical Federation Congress each year.

- Publications – ISEC publishes a monthly e-Newsletter, its yearly study reports and an annual technical journal [CLIMB] to help spread information about space elevators. In addition, there is a magazine filled with space elevator literature called Via Ad Astra.

- Reference material – ISEC is building a Space Elevator Library, including a reference database of Space Elevator related papers and publications. (see section before this on references)

- Outreach – People need to be made aware of the idea of a space elevator. Our outreach activity is responsible for providing the blueprint to reach societal, governmental, educational, and media institutions and expose them to the benefits of space elevators. ISEC members are readily available to speak at conferences and other public events in support of the space elevator. In addition to our monthly e--Newsletter, we are also on Facebook, Linked In, and Twitter.

- Legal – The space elevator is going to break new legal ground. Existing space treaties may need to be amended. New treaties may be needed. International cooperation must be sought. Insurability will be a requirement. Legal activities encompass the legal environment of a space elevator - international maritime, air, and space law. Also, there will be interest within intellectual property, liability, and commerce law. Starting work on the legal foundation well in advance will result in a more rational product.

- History Committee – ISEC supports a small group of volunteers to document the history of space elevators. The committee's purpose is to provide insight into the progress being achieved currently and over the last century.

- Research Committee – ISEC is gathering the insight of researchers from around the world with respect to the future of space elevators. As scientific papers, reports and books are published, the research committee is pulling together this relative progress to assist academia and industry to progress towards an operational space elevator infrastructure.

- Competitions – ISEC has a history of actively supporting competitions that push technologies in the area of space elevators. The initial activities were centered on NASA's Centennial Challenges called "Elevator: 2010." Inside this were two specific challenges: Tether Challenge and Beam Power Challenge. The highlight came when Laser Motive won $900,000 in 2009, as they reached one kilometer in altitude racing other teams up a tether suspended from a helicopter. There were also multiple competitions where different strengths of materials were tested going for a NASA prize – with no winners. In addition, ISEC supports the educational efforts of various organizations, such as the LEGO space elevator climb competition at our Seattle conference. Competitions have also been conducted in both Japan, Israel, and Europe.

ISEC is a traditional not-for-profit 501 (c) (3) organization with a board of directors and four officers: President, Vice President, Treasurer, and Secretary. inbox@isec.org / www.isec.org

www.ingramcontent.com/pod-product-compliance
Lightning Source LLC
Chambersburg PA
CBHW081154180526
45170CB00006B/2080